New Advances in Low-Energy Processes for Geo-Energy Development

New Advances in Low-Energy Processes for Geo-Energy Development

Guest Editors

Daoyi Zhu
Yiqun Zhang
Xiankang Xin
Shuda Zhao
Hongbin Yang

MDPI

Basel • Beijing • Wuhan • Barcelona • Belgrade • Novi Sad • Cluj • Manchester

Guest Editors
Daoyi Zhu
Faculty of Petroleum
China University of
Petroleum-Beijing at Karamay
Karamay
China

Yiqun Zhang
College of Safety and
Ocean Engineering
China University of
Petroleum-Beijing
Beijing
China

Xiankang Xin
School of Petroleum
Engineering
Yangtze University
Wuhan
China

Shuda Zhao
School of Earth Sciences
and Engineering
Xi'an Shiyou University
Xi'an
China

Hongbin Yang
College of Petroleum
Engineering
China University of
Petroleum (Huadong)
Qingdao
China

Editorial Office
MDPI AG
Grosspeteranlage 5
4052 Basel, Switzerland

This is a reprint of the Special Issue, published open access by the journal *Energies* (ISSN 1996-1073), freely accessible at: https://www.mdpi.com/journal/energies/special_issues/7598GICH5H.

For citation purposes, cite each article independently as indicated on the article page online and as indicated below:

Lastname, A.A.; Lastname, B.B. Article Title. *Journal Name* **Year**, *Volume Number*, Page Range.

ISBN 978-3-7258-4147-9 (Hbk)
ISBN 978-3-7258-4148-6 (PDF)
https://doi.org/10.3390/books978-3-7258-4148-6

Contents

About the Editors

Daoyi Zhu

Dr. Daoyi Zhu holds a Bachelor of Science in Applied Chemistry from Yangtze University (2011), a Master's degree in Chemical Engineering (2014), and a Ph.D. in Oil & Gas Field Development Engineering (2018) from China University of Petroleum (Beijing). His academic career includes a visiting scholar appointment at Missouri University of Science and Technology's Department of Petroleum Engineering (2016–2018), and since 2021 he has served as an Associate Professor of Petroleum Engineering at China University of Petroleum-Beijing at Karamay. As an active researcher in energy engineering, Dr. Zhu specializes in oilfield chemistry, reservoir conformance control, CCUS technologies, and geo-energy development. He has made significant scholarly contributions, including publishing over 40 peer-reviewed SCI journal papers, authoring four textbooks (Oilfield Chemistry, Literature Search and Technical Writing, CO2 and Geo-Energy Engineering, and Petroleum Reservoir Engineering), and translating multiple technical works. His editorial leadership extends to roles as Guest Editor and Executive Board Member for prominent international journals in the field.

Yiqun Zhang

Yiqun Zhang is a Professor and Head of the Department of Offshore Oil & Gas Engineering at China University of Petroleum-Beijing (CUP). He graduated from CUP with a Bachelor of Science degree in Petroleum Engineering in 2011 and received a Ph.D. in Petroleum Engineering from Heriot-Watt University, Edinburgh, in 2015. He was awarded the ARMA Future Leader, Beijing Nova Program, and the Excellent Young Scientists Fund grant by the National Natural Science Foundation of China (NSFC). His research interest focuses on drilling and completion engineering and applied fluid mechanics, including the development of natural gas hydrate and geothermal resources.

Xiankang Xin

Dr. Xiankang Xin graduated from Yangtze University with a Bachelor's degree in petroleum engineering in 2009, a Master's degree in Oil & Gas Field Development Engineering in 2015, and a Ph.D. in Oil & Gas Field Development Engineering from China University of Petroleum (Beijing) in 2019. In 2019, he joined the School of Petroleum Engineering of Yangtze University. He is currently an Associate Professor of Petroleum Engineering (since 2022). Dr. Xin focuses on reservoir engineering, numerical reservoir simulation, and enhanced oil recovery. He has published more than 20 peer-reviewed journal articles.

Shuda Zhao

Dr. Shuda Zhao graduated from Missouri University of Science and Technology with a Bachelor's degree in Geology and Geoscience in 2017. He continued his studies at the same institution and earned a Ph.D. in Petroleum Engineering in 2022. In 2023, he joined Xi'an Shiyou University as a Lecturer in the Department of Petroleum Geology at the School of Earth Sciences and Engineering. His research focuses on oil and gas field development, conformance control, and enhanced oil recovery (EOR). Dr. Zhao has published three peer-reviewed journal articles and co-authored one academic book in his field.

Hongbin Yang

Dr. Hongbin Yang graduated from China University of Petroleum (East China) with a Bachelor's degree in Petroleum Engineering in 2010 and obtained his Master's and Ph.D. degrees in Oil & Gas Field Development Engineering from China University of Petroleum (East China) and China University of Petroleum (Beijing) in 2013 and 2016, respectively. He conducted research as a postdoctoral teacher at the Petroleum and Natural Gas Engineering Postdoctoral Station of China University of Petroleum (East China) from August 2016 to July 2018. From July 2018 to December 2021, he served as a lecturer at the university and as an Associate Professor of the college (since 2021). Dr. Yang is mainly engaged in the scientific research work of oilfield chemistry and enhanced oil recovery, CCUS and unconventional geological energy development, and has been awarded the Excellent Youth Project of Shandong Provincial Natural Science Foundation based on the relevant research results. In addition, he has published 49 SCI papers as the first or corresponding author, authored two academic works titled Theory and Technology of Polymer Microsphere Deep Profile Control and Flooding in Complex Reservoir and Theory and Technology of Amphiphilic Polymer Flooding in High-Salinity Oil Reservoirs, and served as a young editor for domestic and foreign academic journals such as Petroleum Science, Journal of Northeast Petroleum University, and Journal of Xi'an Shiyou University.

energies

Editorial

New Advances in Low-Energy Processes for Geo-Energy Development

Daoyi Zhu

Faculty of Petroleum, China University of Petroleum-Beijing at Karamay, Karamay 834000, China; zhudaoyi@cupk.edu.cn

Abstract: The development of geo-energy resources, including oil, gas, and geothermal reservoirs, is being transformed through the creation of low-energy processes and innovative technologies. This Special Issue compiles cutting-edge research aimed at enhancing efficiency, sustainability, and recovery during geo-energy extraction. The published studies explore a diverse range of methodologies, such as the nanofluidic analysis of shale oil phase transitions, deep electrical resistivity tomography for geothermal exploration, and hybrid AI-driven production prediction models. Their key themes include hydraulic fracturing optimization, CO_2 injection dynamics, geothermal reservoir simulation, and competitive gas–water adsorption in ultra-deep reservoirs, and these studies combine advanced numerical modeling, experimental techniques, and field applications to address challenges in unconventional reservoirs, geothermal energy exploitation, and enhanced oil recovery. By bridging theoretical insights with practical engineering solutions, this Special Issue provides a comprehensive foundation for future innovations in low-energy geo-energy development.

Keywords: low-energy processes; hydraulic fracturing; shale oil recovery; geothermal reservoir simulation; CO_2 injection; artificial intelligence; nanofluidic analysis; tight sandstone gas; enhanced geothermal systems

1. Introduction

Geo-energy resources, which include conventional fossil fuels (e.g., oil and natural gas) and unconventional resources (e.g., geothermal, shale gas, and tight oil), remain indispensable to global energy systems [1–3]. These resources are broadly classified as either extractive (hydrocarbons) and renewable (geothermal) [4–6], with each playing a critical role in meeting energy demands and balancing environmental sustainability [7,8]. With industrialization and economic recovery driving energy consumption, optimizing the extraction and utilization of these resources through low-energy technologies has become a pivotal focus of current research [9–11].

In conventional hydrocarbon development, advancements in drilling, completion, and enhanced oil recovery (EOR) technologies have significantly improved the efficiency of these processes and reduced their energy intensity [12–14]. Innovations such as intelligent well systems, high-performance fracturing fluids, and CO_2 injection techniques are reshaping the stimulation and production of reservoirs [15,16]. For instance, polymer flooding and steam-assisted methods are now being adapted to unconventional reservoirs in order to address challenges like low permeability and reservoir heterogeneity [17,18]. Meanwhile, numerical modeling [19–21] and big data analytics [22–24] are enabling precise reservoir management, minimizing the energy wasted during extraction.

The exploitation of unconventional geo-energy resources, such as geothermal and shale gas, demands even greater innovation [25,26]. Geothermal systems, vital for achieving carbon neutrality, require better heat extraction methods, including better approaches to reservoir fracturing and advanced well designs [8]. Similarly, shale gas and tight oil reservoirs benefit from low-energy hydraulic fracturing, gas–water flow optimization, and the in situ upgrading of their technology [27,28]. Emerging approaches like electromagnetic heating and solar–thermal integration further underscore the shift toward sustainable extraction [29–31].

This collection, which is in conversation with the Special Issue of *Energies*, places an emphasis on fundamental innovations and has compiled 10 new publications on the original application of new ideas and on methodologies that will lead to new advances in low-energy processes for geo-energy development.

2. Review of the Research Presented in This Special Issue

The papers featured in this Special Issue showcase state-of-the-art advancements in low-energy technologies used for geo-energy development. These innovations span three critical areas: unconventional hydrocarbon recovery, geothermal resource exploitation, and intelligent simulation techniques that enhance oil and gas production while minimizing energy consumption.

2.1. Unconventional Hydrocarbon Development Using Low-Energy Technologies

Unconventional reservoirs, such as shale reservoirs and tight formations, require innovative approaches to minimize energy consumption while maximizing recovery. Several studies in this Special Issue address critical challenges in hydraulic fracturing, phase behavior, and gas–water interactions. To study the behavior of fluids in shale oil reservoirs at the nanoscale, Lu et al. [32] investigated wax precipitation dynamics in shale oil reservoirs during hydraulic fracturing. Their nanofluidic experiments revealed that the fracturing fluid alters phase transitions, with wax accumulating in microcracks and reducing gas mobility. This insight helps optimize the design of fracturing fluids to mitigate formation damage. Meanwhile, Hu et al. [33] advanced our understanding of CO_2 injection dynamics by visualizing nanoscale salt precipitation in shale pores, which revealed dissolution rates slower than classical theory predicts—a critical insight for mitigating formation damage in CO_2-enhanced oil recovery (EOR) projects.

Jia et al. [34] developed a semi-analytical model of the gas–water flow in hydraulically fractured shale that incorporates water invasion effects. Their findings suggest that residual fracturing fluids hinder gas production, highlighting the need for improved flowback strategies. Liu et al. [35] used molecular dynamics to study CH_4 and H_2O interactions in ultra-deep tight sandstone. Their results show that water adsorption significantly restricts the flow of gas, particularly at high water saturations, providing guidance for production optimization in water-bearing formations.

2.2. Geothermal Resource Exploitation Using Low-Energy Technologies

Efficient geothermal energy extraction requires advanced reservoir characterization and stimulation techniques and optimized heat extraction methods. This Special Issue highlights innovative approaches to improving single-well heating efficiency and the performance of enhanced geothermal systems (EGSs). Sáez Blázquez et al. [36] demonstrated the effectiveness of geoelectrical methods in identifying medium-to-low-enthalpy geothermal resources. Their 2D/3D subsurface models revealed fractured carbonate formations with high geothermal potential, significantly reducing the uncertainty of exploration. Zhu et al. [37] conducted simulations using the TOUGH2-BIOT program to analyze EGS reservoir stimulation and compare heat extraction methods. Their results show that horizontal wells achieve higher heat recovery ratios than traditional double-well systems and double-pipe heat exchange configurations, providing a more sustainable solution for long-term geothermal exploitation. Feng et al. [38] evaluated a closed-loop single-well system in Northeast China. By optimizing key injection parameters, they maximized the system's heat extraction efficiency. Their simulations confirmed the system's viability for district heating applications in low-temperature environments. Additionally, TOUGHREACT modeling revealed that reducing the injection flow rate, injection temperature, and Mg^{2+}/K^{+} concentrations in the injected water can suppress secondary mineral formation and delay near-well reservoir clogging, further improving the system's longevity [39].

2.3. Intelligent Technologies and Numerical Simulation Innovations for Low-Energy Oil and Gas Development

Artificial intelligence (AI) and advanced numerical modeling are revolutionizing low-energy geo-energy development by enabling more efficient and accurate optimization strategies. Kong et al. [40] developed a hybrid two-stage decomposition–LSTM model for oil production forecasting in complex volcanic reservoirs. Their approach significantly outperforms conventional methods, delivering higher-precision predictions to support data-driven reservoir management. When applied to the Jinlong oilfield, optimized injection–production parameters reduced inefficient fluid cycling, enhancing the efficiency of oil recovery. Liu et al. [41] addressed the challenge of limited early-time pressure data in tight reservoirs by developing a physics-constrained workflow that enhances permeability and skin factor estimation, thereby improving the characterization of tight oil reservoirs. Together, these AI-driven and modeling innovations demonstrate the transformative potential of integrating cutting-edge computational approaches into geo-energy engineering.

3. Conclusions

Many academics from a variety of fields, from natural sciences to engineering, have been researching the recovery of unconventional hydrocarbons, geothermal energy exploitation, and intelligent simulation techniques. New theories and technology are presented in this Special Issue. These new experimental methods, numerical simulation technology, and pilot cases can help readers better understand and be inspired by the cutting-edge technologies being used in the field of low-energy geo-energy development.

Acknowledgments: I sincerely appreciate the invaluable guidance and support provided by the editorial team throughout this process. I would also like to extend my deepest gratitude to the contributing academic editors for their expertise and dedication. Finally, I wish to thank the authors for their outstanding contributions and the anonymous reviewers for their rigorous evaluations and insightful feedback, which were instrumental in shaping this publication.

Conflicts of Interest: The author declares no conflicts of interest.

References

1. Wan, Y.; Yuan, Y.; Zhou, C.; Liu, L. Multiphysics coupling in exploitation and utilization of geo-energy: State-of-the-art and future perspectives. *Adv. Geo-Energy Res.* **2023**, *10*, 7–13.
2. Wang, W.; Wang, X.; Chen, Y.; Liu, C.; Zhang, Y. Unconventional resources: Provenance analysis, sediment transport, reservoir evaluation, geo-energy. *Front. Earth Sci.* **2024**, *12*, 1530200. [CrossRef]
3. Wang, J.; Xie, H.-P.; Matthai, S.K.; Hu, J.-J.; Li, C.-B. The role of natural fracture activation in hydraulic fracturing for deep unconventional geo-energy reservoir stimulation. *Pet. Sci.* **2023**, *20*, 2141–2164. [CrossRef]
4. Zhao, M.; Yuan, B.; Liu, Y.; Zhang, W.; Zhang, X.; Guo, W. Dynamic prediction of fracture propagation in horizontal well hydraulic fracturing: A data-driven approach for geo-energy exploitation. *Geoenergy Sci. Eng.* **2024**, *241*, 213182. [CrossRef]

5. Khaleghi, K.; Livescu, S. A review of vertical closed-loop geothermal heating and cooling systems with an Emphasis on the importance of the subsurface. *J. Pet. Sci. Eng.* **2023**, *220*, 111137. [CrossRef]
6. Darko, C.K.; Liu, Y.; Wei, M.; Bai, B.; Schuman, T. Enhancing geothermal efficiency: Experimental evaluation of a high-temperature preformed particle gel for controlling preferential fluid flow. *Renew. Energy* **2024**, *235*, 121417. [CrossRef]
7. Winterfeld, P.; Bai, B.; Wu, Y.-S. Using Preformed Particle Gels to Control Transport in Geothermal Reservoirs: Mathematical Modeling. In Proceedings of the SPE Reservoir Simulation Conference, Galveston, TX, USA, 25–27 March 2025. [CrossRef]
8. Wang, D.-B.; Qin, H.; Wang, Y.-L.; Hu, J.-Q.; Sun, D.-L.; Yu, B. Experimental study of the temporary plugging capability of diverters to block hydraulic fractures in high-temperature geothermal reservoirs. *Pet. Sci.* **2023**, *20*, 3687–3699.
9. Fassihi, M.R.; Kovscek, A.R. *Low-Energy Processes for Unconventional Oil Recovery*; Society of Petroleum Engineers: Richardson, TX, USA, 2017.
10. Xiong, R.; Guo, J.; Kiyingi, W.; Gao, C.; Wang, L.; Luo, J.; Song, H.; Wang, X. Technical transformation of heavy/ultra-heavy oil production in China driven by low carbon goals: A review. *J. Clean. Prod.* **2024**, *458*, 142531. [CrossRef]
11. Zhang, J.; Lin, H.; Li, S.; Yang, E.; Ding, Y.; Bai, Y.; Zhou, Y. Accurate gas extraction (AGE) under the dual-carbon background: Green low-carbon development pathway and prospect. *J. Clean. Prod.* **2022**, *377*, 134372.
12. Zhu, D.-Y.; Fang, X.-Y.; Sun, R.-X.; Xu, Z.-H.; Liu, Y.; Liu, J.-Y. Development of degradable pre-formed particle gel (DPPG) as temporary plugging agent for petroleum drilling and production. *Pet. Sci.* **2021**, *18*, 479–494. [CrossRef]
13. Zhu, D.; Zhao, Q.; Chen, P.; Lu, J.; Yang, Y.; Guo, S.; Zhang, T. Laboratory Evaluation of Antileakage Performance against CO_2 of Alkali-Activated Gel-Reinforced Cement for Carbon Capture, Utilization, and Storage. *SPE J.* **2025**, 1–16. [CrossRef]
14. Zhu, D.; Qin, J.; Gao, Y.; Guo, S.; Bai, L.; Zhao, Y.; Zhao, Q.; Zhang, H. Black Nanosheet-Enhanced Low-Salinity Water System for Improving Oil Recovery in a Low-Permeability Conglomerate Reservoir. *Energy Fuels* **2023**, *37*, 11893–11901.
15. Bai, L.; Shi, C.; Tang, K.; Xie, H.; Yang, S.; Zhu, D. Study on migration and plugging performance of polymer gel in fractured cores using nuclear magnetic resonance technology. *Geoenergy Sci. Eng.* **2023**, *227*, 211891.
16. Shah, M.S.; Shah, S.N. Comparative assessment of mechanical and chemical fluid diversion techniques during hydraulic fracturing in horizontal wells. *Pet. Sci.* **2023**, *20*, 3582–3597. [CrossRef]
17. Seright, R. Potential for polymer flooding reservoirs with viscous oils. *SPE Reserv. Eval. Eng.* **2010**, *13*, 730–740. [CrossRef]
18. Zhu, D.; Bai, B.; Hou, J. Polymer gel systems for water management in high-temperature petroleum reservoirs: A chemical review. *Energy Fuels* **2017**, *31*, 13063–13087.
19. Wang, Y.-Z.; Cao, R.-Y.; Jia, Z.-H.; Wang, B.-Y.; Ma, M.; Cheng, L.-S. A multi-mechanism numerical simulation model for CO_2-EOR and storage in fractured shale oil reservoirs. *Pet. Sci.* **2024**, *21*, 1814–1828. [CrossRef]
20. Jiang, G.-C.; Sheng, K.-M.; He, Y.-B.; Yang, L.-L.; Dong, T.-F.; Sun, Z.; Jiang, K.-L. Numerical simulation of the temporal and spatial evolution of sandstone pore type reservoir damage types and severity. *Sci. Rep.* **2024**, *14*, 25401.
21. Cai, J.; Qin, X.; Xia, X.; Jiao, X.; Chen, H.; Wang, H.; Xia, Y. Numerical modeling of multiphase flow in porous media considering micro-and nanoscale effects: A comprehensive review. *Gas Sci. Eng.* **2024**, *131*, 205441.
22. Jiao, Y.; Ji, J.; Yang, Y.; Yin, H. Big data and artificial intelligence-based numerical simulation and optimization of microbial enhanced oil recovery: Model development, algorithm application, and prospects. *Adv. Resour. Res.* **2025**, *5*, 551–568.
23. Lawal, A.; Yang, Y.; He, H.; Baisa, N.L. Machine learning in oil and gas exploration: A review. *IEEE Access* **2024**, *12*, 19035–19058.
24. Favour, D.A. Petroleum Industry Value Chain Optimization: The Inevitability of Artificial Intelligence and Data Science in Midstream and Downstream Development. In Proceedings of the SPE Nigeria Annual International Conference and Exhibition, Lagos, Nigeria, 5–7 August 2024. [CrossRef]
25. Li, G.; Song, X.; Shi, Y.; Wang, G.; Huang, Z. Current status and construction scheme of smart geothermal field technology. *Pet. Explor. Dev.* **2024**, *51*, 1035–1048. [CrossRef]
26. Xue, L.; Xu, S.; Nie, J.; Qin, J.; Han, J.-X.; Liu, Y.-T.; Liao, Q.-Z. An efficient data-driven global sensitivity analysis method of shale gas production through convolutional neural network. *Pet. Sci.* **2024**, *21*, 2475–2484. [CrossRef]
27. Lobo, F.L.; Wang, H.; Huggins, T.; Rosenblum, J.; Linden, K.G.; Ren, Z.J. Low-energy hydraulic fracturing wastewater treatment via AC powered electrocoagulation with biochar. *J. Hazard. Mater.* **2016**, *309*, 180–184.
28. Hu, Z.; Wang, H. Feasibility study of energy storage using hydraulic fracturing in shale formations. *Appl. Energy* **2024**, *354*, 122251. [CrossRef]
29. Liao, Q.; Wang, B.; Chen, X.; Tan, P. Reservoir stimulation for unconventional oil and gas resources: Recent advances and future perspectives. *Adv. Geo-Energy Res.* **2024**, *13*, 7–9. [CrossRef]
30. Bera, A.; Babadagli, T. Status of electromagnetic heating for enhanced heavy oil/bitumen recovery and future prospects: A review. *Appl. Energy* **2015**, *151*, 206–226. [CrossRef]
31. Sun, C.; Liu, W.; Wang, B.; Ma, T.; Guo, C. Modeling microwave heating for enhanced shale gas recovery: Fully coupled two-phase flows with heat transfer and electromagnetism in deformable reservoirs. *Appl. Therm. Eng.* **2024**, *248*, 123190.
32. Lu, Z.; Wan, Y.; Xu, L.; Fang, D.; Wu, H.; Zhong, J. Nanofluidic Study of Multiscale Phase Transitions and Wax Precipitation in Shale Oil Reservoirs. *Energies* **2024**, *17*, 2415. [CrossRef]

33. Hu, X.; Wang, J.; Zhang, L.; Xiong, H.; Wang, Z.; Duan, H.; Yao, J.; Sun, H.; Zhang, L.; Song, W.; et al. Direct Visualization of Nanoscale Salt Precipitation and Dissolution Dynamics during CO_2 Injection. *Energies* **2022**, *15*, 9567. [CrossRef]
34. Jia, P.; Niu, L.; Li, Y.; Feng, H. A Practical Model for Gas–Water Two-Phase Flow and Fracture Parameter Estimation in Shale. *Energies* **2023**, *16*, 5140. [CrossRef]
35. Liu, Y.; Pei, X.; Yang, F.; Zhong, J.; Dai, L.; Wang, C.; Zhou, T.; Li, Y.; Xiao, S. Molecular Simulation Study of Gas–Water Adsorption Behavior and Mobility Evaluation in Ultra-Deep, High-Pressure Fractured Tight Sandstone Reservoirs. *Energies* **2025**, *18*, 2175. [CrossRef]
36. Sáez Blázquez, C.; Martín Nieto, I.; Carrasco, J.; Carrasco, P.; Porras, D.; Maté-González, M.Á.; Farfán Martín, A.; González-Aguilera, D. Applying Deep Electrical-Resistivity Tomography Techniques for the Exploration of Medium- and Low-Geothermal Energy Resources. *Energies* **2024**, *17*, 1836. [CrossRef]
37. Zhu, J.; Cui, Z.; Feng, B.; Ren, H.; Liu, X. Numerical Simulation of Geothermal Reservoir Reconstruction and Heat Extraction System Productivity Evaluation. *Energies* **2023**, *16*, 127. [CrossRef]
38. Feng, B.; Ren, H.; Yang, Y.; Cui, Z.; Zhao, J. Comparative Analysis of Heating Efficiency of a Single-Well Geothermal System in the Cold Region of Northeast China. *Energies* **2023**, *16*, 1884. [CrossRef]
39. Feng, B.; Cui, Z.; Liu, X.; Shangguan, S.; Qi, X.; Li, S. Effects of Water–Rock Interaction on the Permeability of the Near-Well Reservoir in an Enhanced Geothermal System. *Energies* **2022**, *15*, 8820. [CrossRef]
40. Kong, X.; Liu, Y.; Xue, L.; Li, G.; Zhu, D. A Hybrid Oil Production Prediction Model Based on Artificial Intelligence Technology. *Energies* **2023**, *16*, 1027. [CrossRef]
41. Liu, T.; Jiang, L.; Liu, J.; Ni, J.; Liu, X.; Diwu, P. A Novel Workflow for Early Time Transient Pressure Data Interpretation in Tight Oil Reservoirs with Physical Constraints. *Energies* **2023**, *16*, 245. [CrossRef]

energies

Article

Nanofluidic Study of Multiscale Phase Transitions and Wax Precipitation in Shale Oil Reservoirs

Zhiyong Lu [1], Yunqiang Wan [2,*], Lilong Xu [3], Dongliang Fang [1], Hua Wu [2] and Junjie Zhong [3,*]

[1] Jianghan Oilfield Branch of Sinopec Group, Wuhan 430223, China; 45963ilymf@gmail.com (Z.L.); 1802010613@s.upc.edu.cn (D.F.)
[2] Sinopec Shale Oil and Gas Exploration and Development Key Laboratory, Exploration and Development Research Institute, Sinopec Jianghan Oilfield Company, Wuhan 430223, China; bz23020007@s.upc.edu.cn
[3] State Key Laboratory of Deep Oil and Gas, China University of Petroleum (East China), Qingdao 266580, China; lilongxu2020@163.com
* Correspondence: wanyq.jhyt@sinopec.com (Y.W.); zhongjunjie@upc.edu.cn (J.Z.)

Abstract: During hydraulic fracturing of waxy shale oil reservoirs, the presence of fracturing fluid can influence the phase behavior of the fluid within the reservoir, and heat exchange between the fluids causes wax precipitation that impacts reservoir development. To investigate multiscale fluid phase transition and microscale flow impacted by fracturing fluid injection, this study conducted no-water phase behavior experiments, water injection wax precipitation experiments, and water-condition phase behavior experiments using a nanofluidic chip model. The results show that in the no-water phase experiment, the gasification occurred first in the large cracks, while the matrix throat was the last, and the bubble point pressure difference between the two was 12.1 MPa. The wax precipitation phenomena during fracturing fluid injection can be divided into granular wax in cracks, flake wax in cracks, and wax precipitation in the matrix throat, and the wax mainly accumulated in the microcracks and remained in the form of particles. Compared with the no-water conditions, the large cracks and matrix throat bubble point in the water conditions decreased by 6.1 MPa and 3.5 MPa, respectively, and the presence of the water phase reduced the material occupancy ratio at each pore scale. For the smallest matrix throat, the final gas occupancy ratio under the water conditions decreased from 32% to 24% in the experiment without water. This study provides valuable insight into reservoir fracture modification and guidance for the efficient development of similar reservoirs.

Keywords: shale oil reservoir; hydraulic fracturing; nanofluidic; phase behavior experiment; wax precipitation

1. Introduction

The demands for oil and natural gas are increasing as human society continues to develop. However, conventional oil and gas resources are becoming scarce. As a result, people are now focusing on unconventional resources, such as shale oil and gas, which have shown great potential [1–4]. Shale reservoirs are extremely dense reservoirs characterized by high heterogeneity, ultralow porosity and permeability [5–7], and abundant nanopores [8,9]. Intermolecular interaction forces between the confined fluid and the pore wall in nanoscale space become as important as intermolecular interactions within the confined fluid, and the strong solid–liquid interactions have a significant impact on the adsorption, distribution, mass transfer, and thermodynamic properties of the fluid. These interactions also cause noticeable differences in the physical properties of the fluid in both the bulk phase and nanopores [10,11].

The study of phase behavior in confined spaces has been explored extensively using experimental [12–14], analytical [15,16], and simulation methods [17–19]. Xu et al. [20] conducted a comprehensive study on the two-phase transport properties of shale gas and water in hydrophilic and hydrophobic nanopores by combining molecular dynamics simulations and analytical modeling. Zhang et al. [21] developed a comprehensive thermodynamic model to study the phase behavior of nanopore-confined C1–C10 and C1–C4 mixtures based on pore geometry. The results show that capillarity and adsorption lead to significant deviations in the phase envelope. To study the effect of dynamic capillary pressure on two-phase flow processes, Abbasi et al. [22] proposed a hidden iteration method and implemented it. The results show that the algorithm has good convergence and, considering the dynamic capillary effect, is the key to simulating two-phase flow in porous media. Li et al. [23] proposed a two-phase flow analysis model that takes into account the presence of the water phase. The results show that the oil flow curves of the two-phase and single-phase models are both parabolic, but because of the influences of the flow space and internal water, the oil flow rate decreases. Sharma [24] studied the confined phase behavior of ethane at different pore sizes and temperatures in the ordered mesoporous material MCM-41 at the nanoscale. The results show that the capillary condensation pressure increases with an increasing pore size and temperature.

Shale oil is a complex mixture that comprises waxes, colloids, asphaltenes, aromatics, and light hydrocarbons. This intricate composition poses a significant challenge for the development of shale oil, particularly when the reservoir fluids have a high wax content [25,26]. The shale oil and gas resources of Liang2's lower submember in the Fuxing area are 656 million tons of oil and 530 billion square meters of gas, which are large-scale resources. However, the shale reservoir has a high-waxy fluid, the size difference of matrix nanopores and microfractures is large, and the characterization of the wax precipitation in the reservoir fluid during fracturing will seriously affect the reservoir development effect.

The nanofluidic chip is an emerging experimental technology that allows for the study of heat–mass transfer and physicochemical properties of fluids at the nanoscale. This technology combines semiconductor chip processing with interdisciplinary approaches. It offers precise pore processing at the nanoscale, allows for the design of flow channel structures with a high degree of freedom, enables the flexible treatment of pore surface wettability, and allows for the visual detection and manipulation of fluid states at the nanoscale [27]. It has gradually been applied in the petroleum field, in which Wang et al. applied a nanofluidic device to visualize the phase transitions of pure alkanes and alkane mixtures under nanocondensation as a means of approximating the oil/gas phase behavior in nanopore rocks [28]. Alfi et al. utilized lab-on-a-chip techniques, along with high-resolution imaging, specifically using inverse confocal microscopy equipment, to investigate the phase behavior of hydrocarbons in nanoscale capillaries, also known as nanochannels. The results of this study indicate that this method holds great potential for experimental research on phase behavior within nanoscale pores [29]. Hu et al. utilized nanofluidic technology to construct a shale micromodel to characterize salt precipitation and dissolution. This study successfully differentiated various phases ranging from 50 nm to 5 μm, allowing for the identification of the salt precipitation point and the examination of the precipitation dynamics during CO_2 injection [30].

Because of the unclear understanding of the microscopic seepage mechanism of waxy shale oil at present, conventional laboratory experimental methods cannot reveal the characteristics of continuous phase transition before and after fracturing fluid injection under the condition of a nanoconfined effect, as well as the influence of the wax precipitation on the reservoir fluid flow characteristics. This study carried out a multiscale phase behavior experiment simulation of shale oil revisors after fracturing reconstruction based on nanofluidic technology. The multiscale phase transition phenomenon and wax precipitation phenomenon during water injection was directly observed in the visible area of the experiment, and the influence of the existence of the water phase on the multiscale phase transition was revealed through the analysis of the experimental data. This research offers valuable insight into waxy oil reservoir fracturing, identifies potential strategies for the development of analogous shale reservoirs, and serves as a theoretical guide and reference.

2. Experimental Part

2.1. Chip Design and Preparation

According to the reservoir's physical parameters and fracture characteristics with large differences in the pore sizes and scales, the size of the matrix and pore network was determined by considering the contribution degree of the porosity. Subsequently, a chip model that aligns with the conditions of the fractured reservoir was designed and prepared. As shown in Figure 1, the white part is the matrix throat (13 nm), the yellow parts are the matrix pores (510 nm), the red area is the large crack (10 μm level), and the green areas are the microcracks (1 μm level).

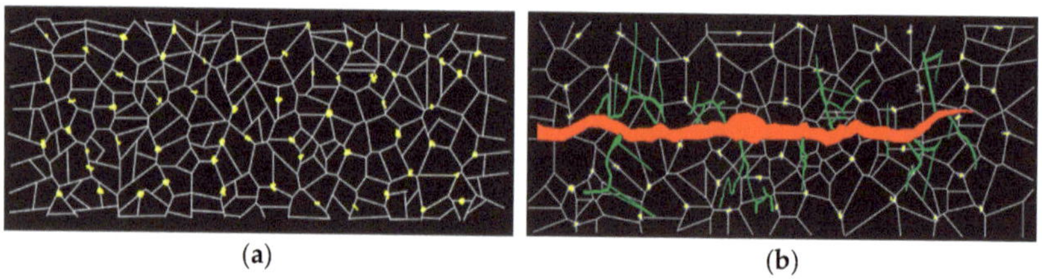

Figure 1. Chip model design. (**a**) The pore network model; (**b**) The pore-crack coupling model.

To fulfill the requirements of the nanofluidic visualization experiments, the nanochannels etched onto the chip were standardized into one-dimensional confined flow channels (specifically, channels with a one-dimensional flow channel section within the designated nanometer scale) during the chip's preparation. The whole chip production included coating and drying, exposure and development, and then the etching machine etched the developed silicon wafer; after the detection met the design requirements, the film was produced, and, finally, the full chip was obtained by anodic bonding [31,32]. The final design of the real chip model is shown in Figure 2.

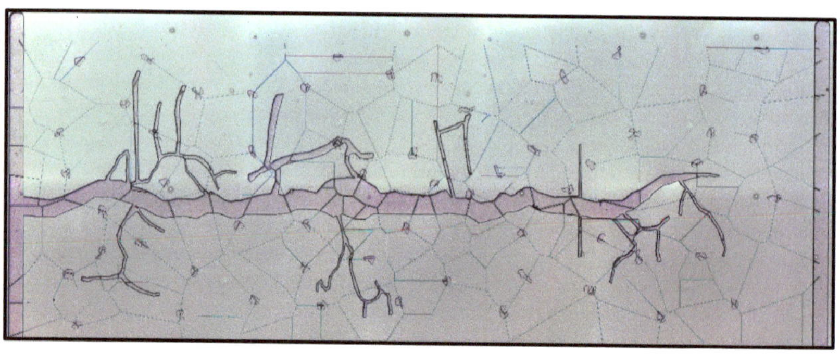

Figure 2. The actual chip models.

2.2. Experimental Materials and Equipment

The oil used in the experiment was taken from the Xingye L1HF well oil sample (Figure 3a). The fracturing fluid was purified water. Based on the chip model, this study built a flow visualization test bench, including a chip fixture, microinjection pump, pressure sensor, constant-temperature water bath, temperature sensor, microscope, and camera, according to the needs of the experiments, as shown in Figure 3b. The chip model and the experimental platform together constitute the whole system used in the conduction of the research experiment.

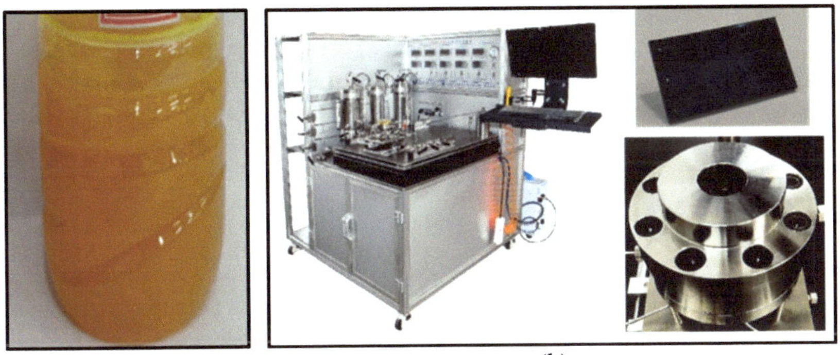

(a) (b)

Figure 3. Images of the related equipment. (**a**) Experimental oil sample; (**b**) Experimental platform.

2.3. Experimental Scheme and Process

The experimental procedure is illustrated in Figure 4. The nanofluidic chip was affixed within a custom-made, high-temperature and high-pressure stainless-steel fixture, designed for conducting tests under elevated temperature and pressure conditions. The fixture was linked to a piston, plunger pump, and gas cylinder to facilitate fluid injection. For simulating the reservoir temperature, a water bath was connected to a hose on the fixture's side, allowing for the circulation of hot water, and the experimental temperature was monitored by inserting a thermocouple. Then, the experiment was carried out according to the following steps: (1) turn on the heating system of the nanofluidic chip stage and the intermediate container, set the experimental temperature to the reservoir temperature (70.8 °C), and vacuum the equipment for 3 h. (2) Step by step, increase the pressure to the reservoir's pressure condition (37.7 MPa) and stabilize for 24 h. (3) Gradually reduce the pressure and record the phase transition and corresponding pressure of the oil at various scales through a microscope. (4) Restore the chip pressure to the reservoir's pressure condition and the oil sample to the single phase; then, the fracturing fluid at room temperature, and continuously injection for 8 PV (pore volume). (5) Stop the injection of the fracturing fluid so that the chip can be stabilized for 24 h under the condition of the reservoir's temperature and the distribution of the oil, water, and wax at all pore scale levels can be balanced. (6) Repeat the pressure reduction operation, and observe and record the phase transition and corresponding pressures of the oil samples at various scales under water conditions through a microscope.

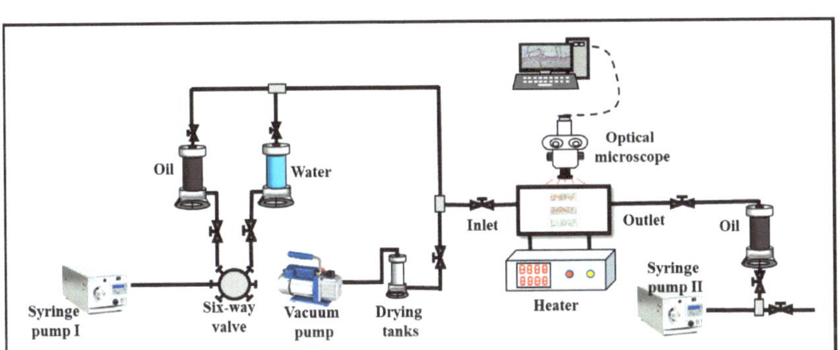

Figure 4. Experiment flowchart.

3. Results and Discussion

After the saturated oil was finished, it was found that at room temperature (25 °C), the oil sample had obvious wax precipitation characteristics, the wax showed a large range of flaked connections, and the darker the color, the more wax was precipitated. Dark-colored areas were more common in the corners of cracks and blind ends. This is because with the continuous erosion of the saturated oil, wax precipitated while being trapped to flow toward the outlet, gradually accumulating in the corners and blind ends (see Figure 5a). Because of the small size of the matrix pores and throats, the amount of wax precipitation was less, and no obvious phenomenon could be seen in the visual field. After the temperature gradually increased to the reservoir's temperature (70.8 °C), the wax in the original precipitation position gradually dissolved in the oil and disappeared from view. A relatively transparent uniform fluid appeared in the visual field, as shown in Figure 5b. It can be seen that the waxy shale oil was in a single liquid phase at the reservoir's temperature.

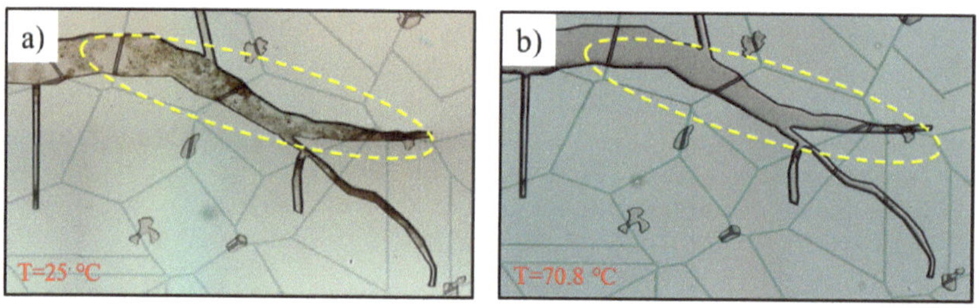

Figure 5. Wax precipitation of the waxy oil under low-temperature and reservoir-temperature conditions. (**a**) Wax precipitation in crack under room temperature; (**b**) Wax disappear under the geological temperature conditions (Yellow dashed circles indicate wax changes at corresponding locations).

3.1. Multiscale Phase Behavior Experiments under No-Water Conditions

After the oil sample in the chip reached a stable equilibrium state, the multiscale phase behavior experiment was carried out under the conditions of no water. Figure 6 shows the bubble points of various pore scales measured under a constant temperature and pressure reduction. It was found that when the pressure was reduced to 34.3 MPa, the large crack was the first to vaporize. This is due to the largest pore size of the large cracks in the chip, and when the pressure is reduced from the inlet, the pressure is transmitted more rapidly to the large cracks so that it first reaches the bubble point. However, the pore sizes of the other channels were relatively small, and they were still in the single-phase oil state due to the space limitation effect. When the pressure continued to decrease to 32.7 MPa, the bubble point was reached, and bubbles appeared in the microcracks. In the subsequent depressurizing process, the matrix pores and throat reached their respective bubble point pressures successively, which were 30.1 MPa and 22.2 MPa, respectively.

The results of the multiscale phase behavior experiments without water show that the smaller the pore size of the channel, the stronger the limiting effect and the more significant the drop in the bubble point pressure. Moreover, with the decrease in pore size, the bubble point has a greater inhibition effect. For example, the bubble point pressures of the microcracks, matrix pores, and throat decreased by 1.6 MPa, 4.2 MPa, and 12.1 MPa, respectively, compared with that of the large cracks, and the degree of reduction gradually increased.

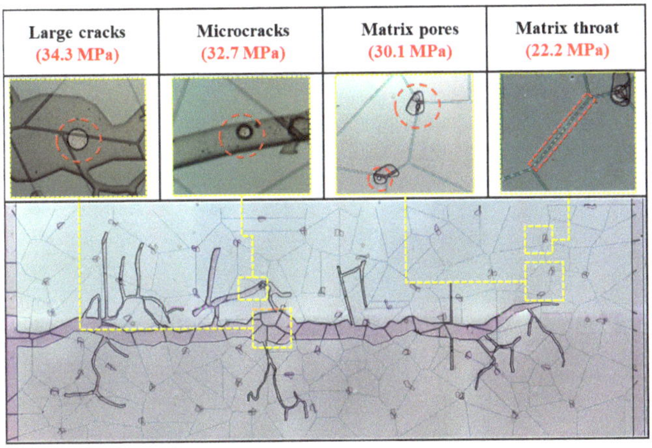

Figure 6. The bubble point pressures, at different scales, under no-water conditions (The yellow dashed box shows where the phase change occurred and enlarges the image, and the red dashed line shows where the bubble is located).

3.2. *Wax Precipitation during Fracturing Fluid Injection*

For the observation of the wax precipitation in the waxy oil during the fracturing fluid injection, the system's pressure was initially restored to the reservoir's pressure conditions. After recovering the oil samples to a single-phase state, water injection at room temperature was initiated to simulate the fracturing fluid injection process, with a total injection volume of 8 PV. The changes in the oil and the wax precipitation process during the fracturing fluid injection were monitored using a camera attached to an optical microscope.

In the process of injection, it was noted that wax crystals gradually developed in the field of view. This is because, with its injection, the fracturing fluid began to exchange heat with the chip and the internal fluid, the continuous injection of the fracturing fluid caused the temperature in the channel to continue to decrease, and the wax previously dissolved in the oil was reprecipitated. At the end of the injection process, the wax precipitation phenomena during the whole experiment can be divided into the following three situations: granular wax in cracks, flake wax in cracks, and wax precipitation in the nanopore throat. The injection time can also be divided into early, middle, and late injections. Figure 7 shows the wax precipitation in the reservoir at different injection volumes of fracturing fluid.

In the early stage of fracturing fluid injection (0~3 PV), the fracturing fluid first played a flooding role, and only oil–water two-phase flow could be observed in the visual field. When the fracturing fluid was injected for 1 PV, because of the small injection amount, the cooling effect of the fluid was not significant, and there was no obvious wax precipitation in the fracture channel. When the injection amount reached 2 PV, a small amount of granular wax began to precipitate and adhere to the wall surface to form a raised surface. When the fracturing fluid was injected for 3 PV, the granular wax gradually increases and accumulates, and a large amount of granular wax can be seen in the whole field of vision, with an average diameter of about 2 microns, and an obvious wax-forming interface appears at the oil–water interface at the blind end of the microfracture. This is because the dead oil at the blind end continued to exchange heat with the injected fracturing fluid, the temperature at the oil–water interface decreased significantly, and the wax could not be taken away, resulting in wax deposition.

In the middle stage of the fracturing fluid injection (4~5 PV), with the continuous erosion of the fracturing fluid, the oil content in the fracture gradually decreased, the granular wax began to migrate and further accumulate, and the flake crystals gradually formed. As can be seen from Figure 7, during the middle stage of the injection, the wax crystals were connected in sheets and showed a pattern of ups and downs on the wall. At this time, it can be seen in the optical microscope field that different shapes of flake crystals developed in different positions of the crack.

In the late stage of the fracturing fluid injection (6~8 PV), the oil in the large cracks and microcracks was removed. With the further erosion of the fracturing fluid, part of the dead oil in the matrix pores and throat was removed, and a small amount of wax occurred in the matrix pores and throat near the fractures. In the visual field, it can be seen that flake wax precipitated from the matrix pores, but the matrix pores and throat were too small to distinguish the crystal shape of the wax phase. It can be observed that the fluid flow in the nanopore throat was less smooth than that before the fracturing fluid injection, and it can be concluded that wax was generated and seriously affected the two-phase oil–water seepage in the matrix throat.

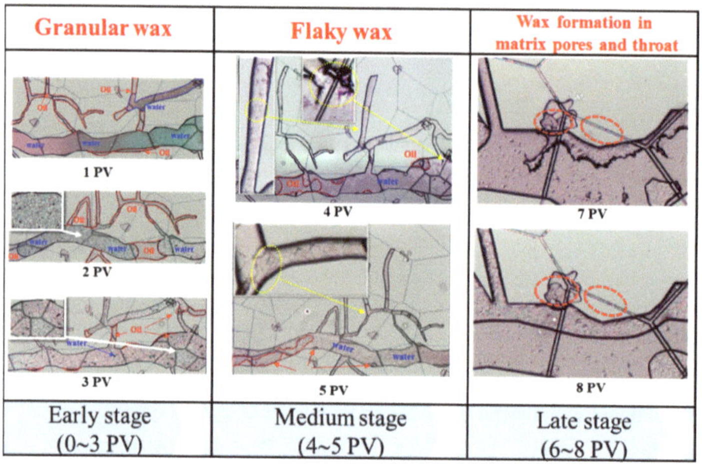

Figure 7. Wax precipitation during the fracturing fluid injection (The dashed and square boxes indicate the wax at the corresponding position).

3.3. Multiscale Phase Behavior Experiments under Water Conditions

After the observation of the wax precipitation phenomenon of the fracturing fluid injection, the injection was stopped, and the temperature was raised to stabilize the chip for 24 h under the condition of the formation temperature so that the distribution of the oil, water, and wax, at all pore scale levels, was stable and balanced. Then, the bubble point pressure, at all pore scales, under water conditions was measured by reducing the pressure again, and the phase behavior and corresponding pressure of the volatile oil, at all scales, under water conditions were observed and recorded through a microscope.

During the 24 h holding stage at the formation temperature, studies have found that the waxy flake crystals during the fracturing fluid injection stage disappeared, and some of them remained in the fracture in the form of particles and no longer melted, affecting the two-phase oil–water seepage flow and causing certain damage to the fracture conductivity. Therefore, it can be proved that for waxy shale oil, the formation of the wax precipitation phenomenon will cause cold damage to the reservoir.

The subsequent pressure reduction experiment results show that under the water conditions, the volatilization of oil still occurs first in the large fracture, and bubbles first nucleate in the water phase at the water–oil interface and then immediately jump to the oil phase to grow and expand. At this time, the bubble point pressure measured in the large fracture was 28.2 MPa, which significantly decreased by 6.1 MPa compared with that without water. When the pressure was reduced to 26.5 MPa, bubbles began to appear in the microcracks. The bubble point pressure of the microcracks was measured. Shortly after bubbles appeared in the microcracks, the bubble point pressure also reached the matrix pores, at which time the pressure was 24.9 MPa. Finally, gas was precipitated in the matrix throat when the pressure dropped to 18.7 MPa, as shown in Figure 8.

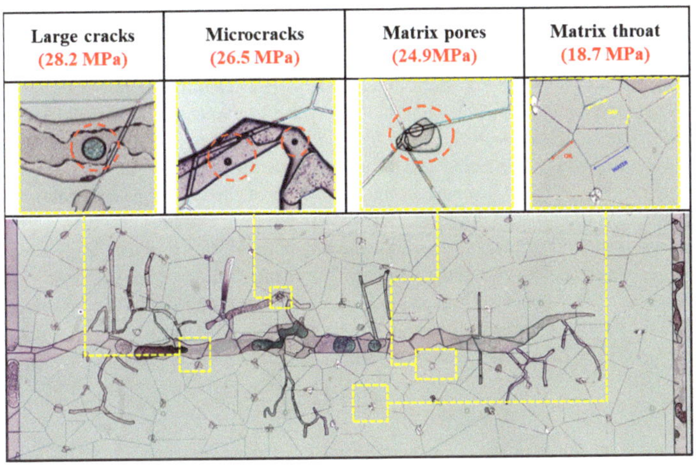

Figure 8. Bubble point pressures, at all pore scales, under water conditions (The yellow dashed box shows where the phase change occurred and enlarges the image, and the red dashed line shows where the bubble is located).

The results of the experiments under the water conditions show that the presence of fracturing fluid can significantly affect the bubble point change, at all pore scales; the presence of the water phase severely inhibits the bubble point; and the bubble point hysteresis occurs at all pore scales. As shown in Figure 9, the larger the pore scale, the more significant the impact. The bubble point pressure of the large cracks and microcracks at the micron scale decreased by about 6 MPa under the water conditions compared with the no-water conditions. The drop in the bubble point pressure in the 100 nm matrix pores was 5.2 MPa, and the drop in the bubble point pressure in the 10 nm matrix throat was only 3.5 MPa. This is also because the smaller the pore scale, the less fracturing fluid that enters, so the smaller the inhibition effect on it.

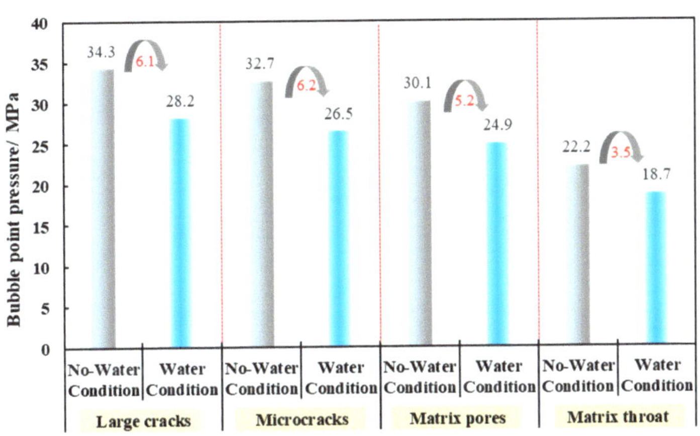

Figure 9. Comparison of the bubble point pressures, at all pore scales, under the no-water and water conditions.

3.4. Material Occupancy Ratios at Different Pore Scales

With micro-nanofluidic technology, the ratio of oil–water/oil–gas–water occupancy in the flow channels at all pore scales in the reservoir can be monitored and captured in real time. Figures 10 and 11 show the proportions of two-phase oil–water and three-phase oil–gas–water, at all pore scales, under the no-water conditions and water conditions. As can be seen in Figure 10, in the experiment at various pore scales under the conditions of no water, the gas occupancy of the large cracks was the highest, which is in line with the normal phenomenon. Because of the high bubble point pressure of the large cracks, gasification occurred first; then, the pressure gradually decreased, and the gas continued to precipitate. As can be seen in the figure, when the bubble point pressure was 34.4 MPa, the gas occupancy ratio was only 0.83%, and then it rose rapidly until the pressure dropped to 15 MPa; the increase flattened gradually, the final gas occupancy ratio reached 78.63%, and the remaining oil is only 21.37%. When the microcracks reached the bubble point, the gas occupancy ratio was 0.37%, and the final gas occupancy ratio was 56.56%. The final gas occupancy ratio of the matrix pores and throat increased to 45.54% and 32%, respectively. It also shows that with the decrease in the pore size, the smaller the gas occupancy ratio, and the more difficult it is for oil to develop in the pores.

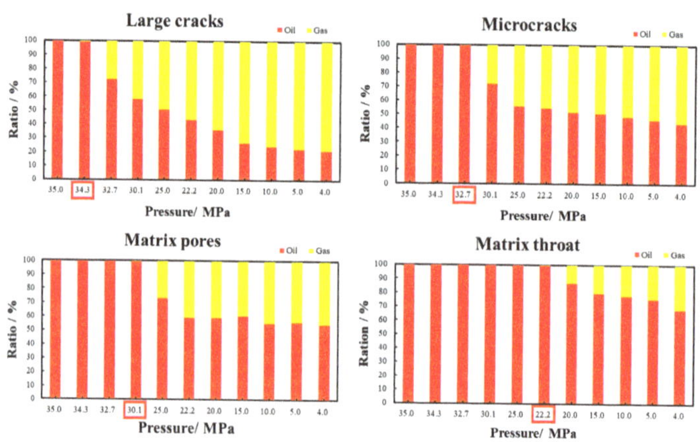

Figure 10. The ratio of oil and gas, at different pore scales, under single-phase conditions (The red box indicates the bubble point at the corresponding scale).

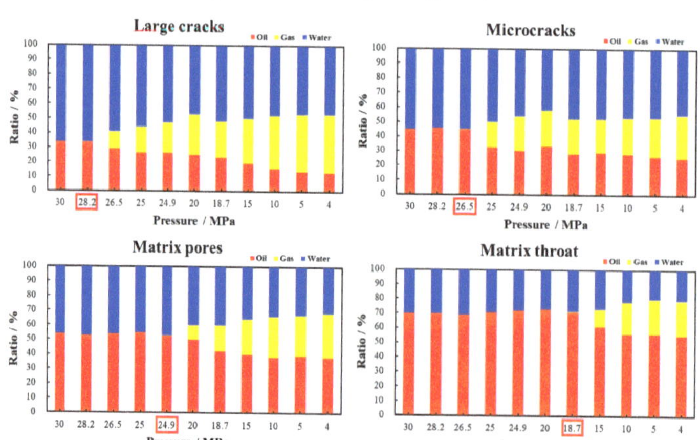

Figure 11. The ratio of oil, gas, and water, at all pore scales, under water conditions (The red box indicates the bubble point at the corresponding scale).

Figure 11 shows the occupancy ratio of the three-phase oil, gas, and water under water conditions. After large-scale fracturing fluid injection, the water phase occupancy of the large cracks reached 66%, that of microcracks was 55%, and the matrix pores and matrix throats were 46% and 30%, respectively. With the gradual reduction in the pressure, the amount of gas precipitation in the pores, at all levels, was inhibited under the water conditions. When the pressure dropped to 4 MPa, the gas occupancy ratios in the pores, at all levels, were 40%, 30%, 30%, and 24%, respectively. The corresponding final oil phase occupancy ratios were 13%, 25%, 38%, and 55%, respectively. The results show that the presence of the water phase inhibited the gas production, at all pore scales, and thus affected the degree of oil phase utilization.

To more intuitively characterize the influence under water conditions on the amount of gas precipitated, at various pore scales, this paper selected the oil–gas/oil–gas–water occupancy ratio for each pore scale under the final experimental pressure of 4 MPa for the calculations. The relative gas–oil occupancy (R_l^g) was used as an index to characterize the degassing of oil samples per unit volume.

$$R_l^g = \frac{R_g}{R_l} \tag{1}$$

where R_g is the gas phase occupancy, at all pore scales, when the pressure is 4 MPa; R_l is the oil phase occupancy, at all pore scales, when pressure is 4 MPa. The corresponding data can be directly read in Figures 10 and 11. Figure 12 shows the trend in the relative gas–oil occupancy, at all pore scales, under the following two conditions: without fracturing fluid and with fracturing fluid. It can be seen from the figure that, affected by the presence of the water phase, the relative gas–oil occupancies in the large cracks, microcracks, matrix pores, and matrix throat decreased by 0.6, 0.1, 0.05, and 0.03, respectively. The decrease was the greatest in the large cracks. It is intuitively proved that gas production, at all pore scales, will be inhibited under water conditions, and the larger the pore scale, the greater the influence of water on the gas precipitation.

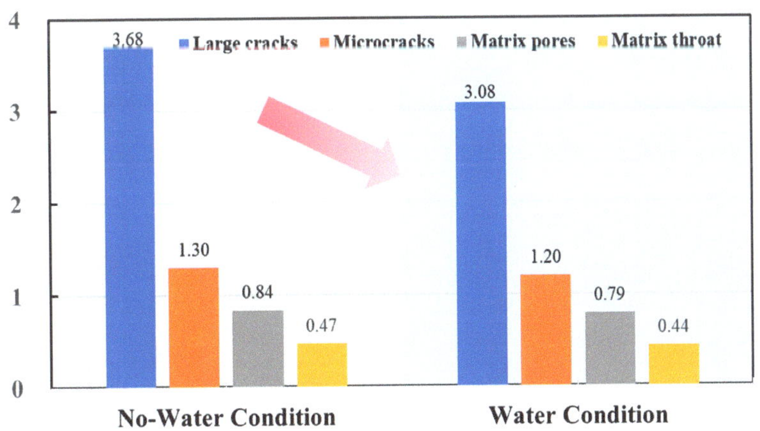

Figure 12. Comparison of the relative gas–oil occupancies, at all pore scales, under the no-water and water conditions.

4. Conclusions

In this work, this study focused on the characteristics of continuous phase transition before and after fracturing fluid injection and the impact of wax precipitation characteristics. Our findings are as follows:

(1) In the multiscale phase behavior experiment under the no-water conditions, it was observed that gasification initiated first in the large cracks, with a corresponding bubble point pressure of 34.3 MPa. In the subsequent process with a constant temperature and pressure reduction, with the decreases in the pore size, bubble point pressure of the microcracks, matrix pores, and matrix throat were 32.7 MPa, 30.1 MPa, and 22.2 MPa, respectively.
(2) During the process of the fracturing fluid injection, the wax precipitation phenomena manifested in the following three cases: granular wax and flake wax in the cracks, and wax formation in the matrix pores and throat. Wax precipitation was primarily concentrated in micron-sized cracks, retaining a granular form. This impeded the fluid flow and, to some extent, compromised the conductivity of the cracks.
(3) The bubble point pressures of the different pore scales decreased further due to the influence of the water phase. The bubble point pressure of the large cracks was 28.2 MPa, representing a significant decrease of 6.1 MPa compared to the no-water conditions, the bubble points corresponding to the microcracks, matrix pores, and matrix throats decreased by 6.2 MPa, 5.2 MPa, and 3.5 MPa, respectively.
(4) The water phase also significantly influenced the material occupancy ratios across the different pore scales. The R_l^g value intuitively proves that gas production at different pore scales will be inhibited under the water conditions, and the larger the pore scale, the greater the influence of the water on the gas precipitation.

The nanofluidic technology was used in this study to reveal the phase transition hysteresis at the multiscale and to illustrate the effects of water. The presence of the water phase would not only precipitate wax in crude oil but also affect the permeability of the pores, to a certain extent, and inhibit the bubble point at each pore scale. Also, the water phase would inhibit the gas precipitation at different pore scales. In the future development of similar waxy shale reservoirs, the impact of water can be reduced as much as possible to enhance oil recovery.

Author Contributions: Methodology, D.F.; Software, D.F.; Formal analysis, Y.W. and H.W.; Investigation, Z.L.; Data curation, Z.L. and L.X.; Writing—original draft, Z.L. and L.X.; Writing—review & editing, Y.W. and J.Z.; Supervision, Y.W. and J.Z.; Funding acquisition, J.Z. All authors have read and agreed to the published version of the manuscript.

Funding: We gratefully acknowledge the generous support from the National Natural Science Foundation of China (No. 42090024, No. 52122402 and No. 52174051), Excellent Young Scholars of Shandong Province (No. 2022HWYQ-072), and Guanghua Scholars of China University of Petroleum (East China) (No. 20210002).

Data Availability Statement: The original contributions presented in the study are included in the article, further inquiries can be directed to the corresponding author.

Conflicts of Interest: Authors Zhiyong Lu and Dongliang Fang were employed by the company Jianghan Oilfield Branch of Sinopec Group. Authors Yunqiang Wan and Hua Wu were employed by the company Exploration and Development Research Institute Sinopec Jianghan Oilfield Company. The remaining authors declare that the research was conducted in the absence of any commercial or financial relationships that could be construed as a potential conflict of interest.

References

1. Chauhan, A.; Salehi, F.; Jalalifar, S.; Clark, S. Two-phase modelling of the effects of pore-throat geometry on enhanced oil recovery. *Appl. Nanosci.* **2021**, *13*, 453–464. [CrossRef]
2. Jia, B.; Su, J. Advancements and Environmental Implications in Oil Shale Exploration and Processing. *Appl. Sci.* **2023**, *13*, 7657. [CrossRef]
3. Sharma, S.; Agrawal, V.; Akondi, R. Role of biogeochemistry in efficient shale oil and gas production. *Fuel* **2020**, *259*, 116247.
4. Taheri-Shakib, J.; Kantzas, A. A comprehensive review of microwave application on the oil shale: Prospects for shale oil production. *Fuel* **2021**, *305*, 121132.
5. Lee, T.; Bocquet, L.; Coasne, B. Activated desorption at heterogeneous interfaces and long-time kinetics of hydrocarbon recovery from nanoporous media. *Nat. Commun.* **2016**, *7*, 11890. [CrossRef]

6. Yang, J.; Hatcherian, J.; Hackley, P.C.; Pomerantz, A.E. Nanoscale geochemical and geomechanical characterization of organic matter in shale. *Nat. Commun.* **2017**, *8*, 2179. [CrossRef] [PubMed]
7. Falk, K.; Coasne, B.; Pellenq, R.; Ulm, F.J.; Bocquet, L. Subcontinuum mass transport of condensed hydrocarbons in nanoporous media. *Nat. Commun.* **2015**, *6*, 6949. [CrossRef] [PubMed]
8. Chen, L.; Jiang, Z.; Jiang, S.; Liu, K.; Yang, W.; Tan, J.; Gao, F. Nanopore Structure and Fractal Characteristics of Lacustrine Shale: Implications for Shale Gas Storage and Production Potential. *Nanomaterials* **2019**, *9*, 390. [CrossRef]
9. Zhang, L.; Li, B.; Jiang, S.; Xiao, D.; Lu, S.; Zhang, Y.; Chen, L. Heterogeneity characterization of the lower Silurian Longmaxi marine shale in the Pengshui area, South China. *Int. J. Coal Geol.* **2018**, *195*, 250–266. [CrossRef]
10. Sobecki, N.; Nieto-Draghi, C.; Di Lella, A.; Ding, D.Y. Phase behavior of hydrocarbons in nano-pores. *Fluid Phase Equilibria* **2019**, *497*, 104–121. [CrossRef]
11. Song, Y.; Song, Z.; Zhang, Z.; Chang, X.; Wang, D.; Hui, G. Phase Behavior of CO_2-CH_4-Water Mixtures in Shale Nanopores Considering Fluid Adsorption and Capillary Pressure. *Ind. Eng. Chem. Res.* **2022**, *61*, 5652–5660. [CrossRef]
12. Ojha, S.P.; Misra, S.; Tinni, A.; Sondergeld, C.; Rai, C. Relative permeability estimates for wolfcamp and eagle ford shale samples from oil, gas and condensate windows using adsorption-desorption measurements. *Fuel* **2017**, *208*, 52–64. [CrossRef]
13. Pini, R.; Benson, S.M. Simultaneous determination of capillary pressure and relative permeability curves from core-flooding experiments with various fluid pairs. *Water Resour. Res.* **2013**, *49*, 3516–3530. [CrossRef]
14. Liu, J.; Sheng, J.J.; Wang, X.; Ge, H.; Yao, E. Experimental study of wettability alteration and spontaneous imbibition in chinese shale oil reservoirs using anionic and nonionic surfactants. *J. Petrol. Sci. Eng.* **2019**, *175*, 624–633. [CrossRef]
15. Miqueu, C.; Grégoire, D. Estimation of pore pressure and phase transitions of water confined in nanopores with non-local density functional theory. *Mol. Phys.* **2020**, *118*, e1742935. [CrossRef]
16. Song, W.; Yao, J.; Li, Y.; Sun, H.; Wang, D.; Yan, X. Gas–water relative permeabilities fractal model in dual-wettability multiscale shale porous media during injected water spontaneous imbibition and flow back process. *Fractals* **2020**, *28*, 2050103. [CrossRef]
17. De Andrade, D.D.C.J.; Nojabaei, B. Phase Behavior and Composition Distribution of Multiphase Hydrocarbon Binary Mixtures in Heterogeneous Nanopores: A Molecular Dynamics Simulation Study. *Nanomaterials* **2021**, *11*, 2431. [CrossRef]
18. Mandal, S.; Singh, S.; Bhattacharjee, S.; Khan, S. Phase behaviour of confined associating fluid in a functionalized slit pore: A Monte Carlo study. *Fluid Phase Equilibria* **2021**, *531*, 112909. [CrossRef]
19. Zhang, T.; Javadpour, F.; Li, J.; Zhao, Y.; Zhang, L.; Li, X. Pore-scale perspective of gas/water two-phase flow in shale. *SPE J.* **2021**, *26*, 828–846. [CrossRef]
20. Xu, H.; Yu, H.; Fan, J.; Zhu, Y.; Wang, F.; Wu, H. Two-phase transport characteristic of shale gas and water through hydrophilic and hydrophobic nanopores. *Energy Fuels* **2020**, *34*, 4407–4420. [CrossRef]
21. Zhang, M.; Liu, R.; Hu, J.; Zhang, Y.; Sheng, G. A Pore-Geometry-Based Thermodynamic Model for the Nanoconfined Phase Behavior in Shale Condensate Reservoirs. *Lithosphere* **2022**, *2022*, 1989358. [CrossRef]
22. Abbasi, J.; Ghaedi, M.; Riazi, M. A new numerical approach for investigation of the effects of dynamic capillary pressure in imbibition process. *J. Pet. Sci. Eng.* **2018**, *162*, 44–54. [CrossRef]
23. Li, R.; Chen, Z.; Wu, K.; Hao, X.; Xu, J. An analytical model for water-oil two-phase flow in inorganic nanopores in shale oil reservoirs. *Pet. Sci.* **2021**, *18*, 1776–1787. [CrossRef]
24. Sharma, K.V.; Alloush, R.M.; Piri, M. Confined phase behavior of ethane in nanoporous media: An experimental investigation probing the effects of pore size and temperature. *Microporous Mesoporous Mater.* **2023**, *351*, 112459. [CrossRef]
25. Yang, F.; Dou, Z.; Xia, X.; Liu, D.; Li, C.; Yao, B.; Zhao, Y. Influence of Thermal Treating Temperature on the Performance of EVA Wax Inhibitor for Changqing Shale Oil. *Energy Fuels* **2023**, *37*, 7798–7808. [CrossRef]
26. Liu, B.; Mahlstedt, N.; Horsfield, B.; Tian, S.; Huo, Q.; Wen, Z.; Pan, Z. Phase behavior and GOR evolution using a natural maturity series of lacustrine oil-prone shale: Implications from compositional modelling. *Org. Geochem.* **2023**, *185*, 104675. [CrossRef]
27. Zhong, J.; Wang, Z.; Sun, Z.; Yao, J.; Yang, Y.; Sun, H.; Zhang, L.; Zhang, K. Research advances in microscale fluid characteristics of shale reservoirs based on nanofluidic technology. *Acta Pet. Sin.* **2023**, *44*, 207–222.
28. Wang, L.; Parsa, E.; Gao, Y.; Ok, J.T.; Neeves, K.; Yin, X.; Ozkan, E. Experimental study and modeling of the effect of nanoconfinement on hydrocarbon phase behavior in unconventional reservoirs. In Proceedings of the SPE Western North American and Rocky Mountain Joint Meeting, Denver, CO, USA, 17–18 April 2014; p. 169581.
29. Alfi, M.; Nasrabadi, H.; Banerjee, D. Experimental investigation of confinement effect on phase behavior of hexane, heptane and octane using lab-on-a-chip technology. *Fluid Phase Equilibria* **2016**, *423*, 25–33. [CrossRef]
30. Hu, X.; Wang, J.; Zhang, L.; Xiong, H.; Wang, Z.; Duan, H.; Yao, J.; Sun, H.; Zhang, L.; Song, W.; et al. Direct Visualization of Nanoscale Salt Precipitation and Dissolution Dynamics during CO_2 Injection. *Energies* **2022**, *15*, 9567. [CrossRef]
31. Zhong, J.; Zhao, Y.; Lu, C.; Xu, Y.; Jin, Z.; Mostowfi, F.; Sinton, D. Nanoscale Phase Measurement for the Shale Challenge: Multicomponent Fluids in Multiscale Volumes. *Langmuir* **2018**, *34*, 9927–9935. [CrossRef]
32. Jatukaran, A.; Zhong, J.; Persad, A.H.; Xu, Y.; Mostowfi, F.; Sinton, D. Direct Visualization of Evaporation in a Two-Dimensional Nanoporous Model for Unconventional Natural Gas. *ACS Appl. Nano Mater.* **2018**, *1*, 1332–1338. [CrossRef]

energies

MDPI

Article

Direct Visualization of Nanoscale Salt Precipitation and Dissolution Dynamics during CO_2 Injection

Xinling Hu [1], Jian Wang [1], Liang Zhang [1], Hongli Xiong [1], Zengding Wang [2], Huazheng Duan [2], Jun Yao [2], Hai Sun [2], Lei Zhang [2], Wenhui Song [2] and Junjie Zhong [2,*]

1 Exploration and Development Research Institute Sinopec Jianghan Oilfield Company, Wuhan 430223, China
2 Research Center of Multiphase Flow in Porous Media, School of Petroleum Engineering, China University of Petroleum (East China), Qingdao 266580, China
* Correspondence: zhongjunjie@upc.edu.cn

Abstract: CO_2 injection to enhance shale oil recovery provides a win-win solution to meet the global fuel shortage and realize ultimate carbon neutrality. When shale reservoirs contain high salinity water, CO_2 injection can result in salt precipitation to block the nanometer pores in the shale, causing undesirable formation damage. Understanding salt precipitation and dissolution dynamics at the nanoscale are fundamental to solving this practical challenge. In this work, we developed a shale micromodel to characterize salt precipitation and dissolution based on nanofluidic technology. By directly distinguishing different phases from 50 nm to 5 μm, we identified the salt precipitation sites and precipitation dynamics during the CO_2 injection. For the salt precipitation in the nanometer network, we identified two precipitation stages. The ratio of the precipitation rates for the two stages is ~7.9 times that measured in microporous media, because of the slow water evaporation at the nanoscale. For the salt precipitation in the interconnected micrometer pores, we found that the CO_2 displacement front serves as the salt particle accumulating site. The accumulated salt particles will in turn impede the CO_2 flow. In addition, we also studied the salt dissolution process in the shale micromodel during water injection and found the classical dissolution theory overestimates the dissolution rate by approximately twofold. This work provides valuable pore-scale experimental insight into the salt precipitation and dissolution dynamics involved in shale formation, with the aim to promote the application of CO_2 injection for shale oil recovery.

Keywords: CCUS; CO_2 injection; shale oil; microfluidics; salt precipitation

1. Introduction

The CO_2 injection is a promising innovation to rejuvenate shale oil reservoirs after primary production [1–3]. Given the nanometer pore sizes and ultra-low permeabilities (down to nano-Darcy) of shale reservoirs [4,5], CO_2 is able to improve the oil recovery efficacy by its excellent mixability with oil, leading to oil swelling and oil viscosity reduction [6,7]. However, when the formation rocks consist of a large proportion of halite, the shale formation water often contains high salinity, especially after hydraulic fracturing with massive water injected [8]. A typical example of this type of shale reservoir is the one discovered in the Qianjiang depression of the Jianghan basin in China, which is characterized by its multiple rhythmic layers of salt rock and gypsum [9]. Injecting CO_2 into the shale oil reservoir with high salinity water could result in salt precipitation blocking the nanometer flow pathway in shale, resulting in undesirable formation damage similar to asphaltene precipitation [10].

The salt precipitation in the porous media during CO_2 injection has been widely studied for carbon sequestration and storage in saline aquifers [11,12]. The saline aquifer is featured by its large pore sizes (at 10^1–10^2 μm) [13] and high permeability (at 10^2–10^3 mD) [14]. By using core-flooding with CT scanning, previous studies have pointed out that salt precipitation in the saline aquifer can reduce formation permeability by up to three orders of magnitude [15–17]. Two types of salt precipitation in porous media have been identified, including homogeneous and local salt precipitations. The homogeneous salt precipitation is believed to cause minor damage to permeability as the pore size in the saline aquifer is orders of magnitude larger than the size of the deposited salt particles. The local salt precipitation, however, is regarded as the main reason to block the near-wellbore area and leads to obvious permeability reduction [18]. Developing chemicals to minimize salt precipitation is thus vital to protect reservoir permeability [19].

The shale reservoir is highlighted by its ultra-small pore sizes (mainly at 10^0–10^2 nm). Thus, even the homogeneous salt precipitation could lead to significant porosity and permeability reduction [20]. For example, salt precipitation has been recently found to cause up to ~90% permeability reduction in tight formation rocks [21,22]. In addition, the salt precipitation in nanometer pores has also been identified to reduce the mechanical strength of shale [23]. At the nanometer pore scale, salt precipitation dynamics during CO_2 injection remain largely unknown. The core flooding method is limited at the nanoscale to understand the salt precipitation dynamics during CO_2 injection. Even with the help of CT scanning, characterizing salt precipitation in the nanometer pores of shale samples remains challenging. Therefore, finding an alternative approach to provide nanometer pore-scale insight into the salt precipitation in shale reservoirs is crucial.

Microfluidics has been used to characterize pore-scale salt precipitation dynamics of CO_2 injection in saline aquifers [24,25]. At the nanometer pore scale, researchers have recently applied nanofluidics to study fluid fundamentals in shale reservoirs down to the sub-10 nm scale, including hydrocarbon phase transitions [26,27] and fluid transport properties [28,29]. Fluid behaviors in nanometer pores are often found to deviate from bulk fluid properties [28,30]. For the CO_2 injections into shale oil reservoirs, a previous nanofluidic study has identified immiscible and miscible flooding phenomena in a 60-nm network chip, as well as huff-n-puff mechanisms [31]. Salt precipitation dynamics during CO_2 injections at the nanoscale have not yet been studied with nanofluidics, and remain obscure from the experimental perspective. To fill this gap, we fabricated a multiscale (50 nm, 500 nm and 5 µm) nanofluidic porous chip to study salt precipitation during the CO_2 injection. We identified the preferential precipitation scale during the CO_2 injection. With image analysis, we characterized the salt precipitation dynamics in the nanometer network and pores, and found deviations to the salt precipitation dynamics in the micrometer porous media of saline aquifers. In addition, we performed water injection after CO_2 injection to dissolve salt, and identified the dissolution dynamics in the nanometer network deviating from classical dissolution theory. This work aims to provide the nanometer pore-scale insight of salt precipitation during CO_2 injections into shale oil reservoirs, from a fundamental and experimental aspect.

2. Material and Methods

To provide a nanometer pore-scale physical micromodel of the shale reservoir, we fabricated a nanofluidic chip considering three levels of pore features (Figure 1): (i) a 50-nm deep interconnected network (fabricated as 52.7 ± 4.6 nm for the channel depth and 2 µm for the channel width) to simulate the connected nanometer pores in shale matrix; (ii) a 500-nm deep interconnected network (fabricated as 516.1 ± 9.8 nm for the channel depth and 20 µm for the channel width) to simulate the nanometer natural fractures in shale; (iii) 5-µm deep pores (fabricated as 5.1 ± 0.1 µm for the pore depth and approximately 100–300 µm for the equivalent diameter on the chip surface) to simulate large dissolution pores in shale. For all fabricated features, the depth of the feature is the governing dimension for the fluid transport and salt precipitation. The width of the feature provides visibility to distinguish different phases (CO_2, water and salt) under the microscope. The nanoporous media flank two main microchannels to supply testing fluids (salt solution, pure water and CO_2). The width of the entire nanoporous media is 2 mm. To fabricate the nanofluidic chip, we repeated photolithography and reactive ion etching three times.

After that, we performed anodic bonding and chip dicing. The detailed multiscale chip fabrication methods are delineated in our previous work [32,33].

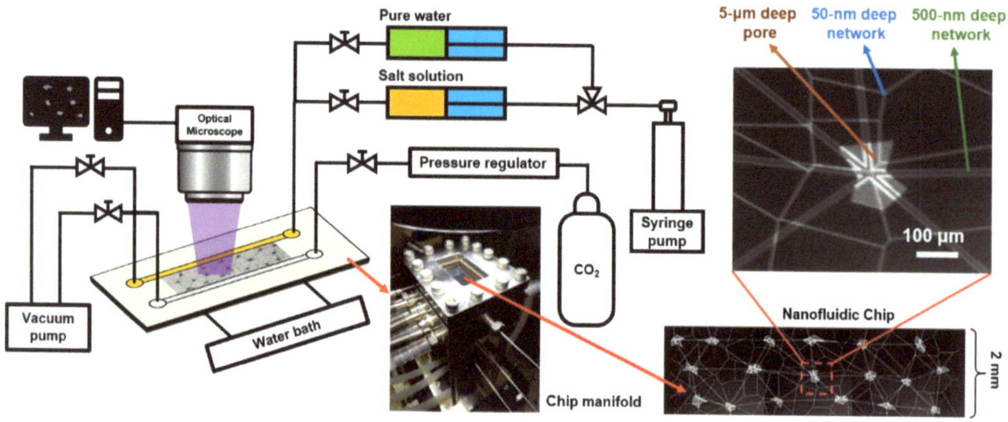

Figure 1. Schematic and photos of the experimental setup and the nanofluidic chip.

The schematic of the experimental setup is shown in Figure 1. The fabricated nanofluidic chip is mounted in a self-manufactured stainless-steel manifold to allow high temperature and pressure tests. The metal manifold is then connected with pistons, syringe pumps and gas cylinders perform the fluid injections. To simulate reservoir temperatures, we connected a water bath to the manifold for circulating hot water, and measured the experimental temperature by inserting thermocouples. In our experiments, the temperature is controlled at 70 ± 1.2 °C for the entire chip, which is close to the shale reservoir temperature. The experimental results are captured as images by the optical microscope connected with a mounted digital camera.

3. Results and Discussion

To study salt precipitation during the CO_2 injection in the nanoporous media. We first vacuumed the chip for 1 h to remove the air initially in the chip. Then, we injected the mixed salt solution (NaCl = 282.9 g/L, Na_2SO_4 = 10.6 g/L, $CaSO_4$ = 1.6 g/L) into the nanofluidic chip. The composition of salts in the mixed solution is formulated based on the salinity analysis results of the produced water in four wells in the field. After salt solution injection, the CO_2 gas is injected at a constant 1 MPa pressure. The pressure is larger than the maximum capillary pressure in the 50-nm deep network (~0.76 MPa). We defined the moment when the gas bubble is observed in the porous media as $t = 0$ s. The salt precipitation results are then recorded by the camera connected with the optical microscope, as shown in Figure 2.

At the early stage of CO_2 injection (~2 s), gas bubbles are generated in the 5-μm deep pores without noticeable salt precipitation (Figure 2a). With the CO_2 further injection, salt contents start to nucleate in the 500-nm deep network (Figure 2b), and then in the 5-μm deep pores (Figure 2c). The precipitated salt then grows and accumulates at both scales with continuous CO_2 injection. However, the salt solution in the 50-nm deep network remains in the liquid phase for the entire CO_2 injection process (Figure 2d).

3.1. Salt Precipitation Dynamics in the Nanometer Network

To quantify the salt precipitation dynamics in the 500-nm deep network, we performed image analysis to extract the 500-nm network from the original images captured by the microscope camera (Figure 3a). The extracted grayscale image sequence is then calculated in MATLAB to quantify the precipitated salt in the 500-nm network. Three stages of the salt precipitation in the 500-nm deep network are identified (Figure 3b). We defined the occupancy of the solid salt particles in the 500-nm deep network as a ratio between the total volume of salt particles and the network (Figure 3a). In the first 1.6 s, CO_2 injection leads to negligible salt deposition, and water displacement and evaporation mainly happen at this stage. After the salt concentration reaches saturation, salt precipitates fast in the 500-nm deep network (1.6 to 4.7 s) due to heterogeneous salt nucleation in the solution. After 4.7 s, salt contents precipitate and deposit slowly in the 500-nm deep network due to water evaporation.

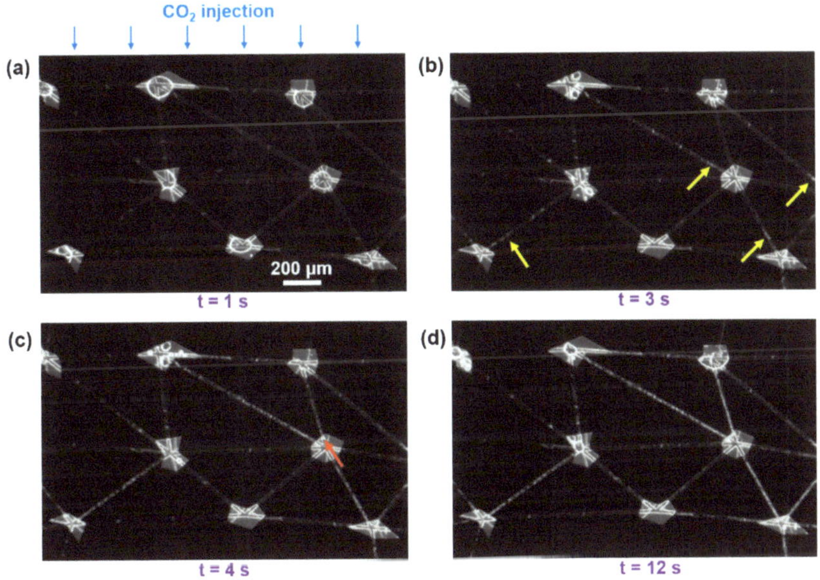

Figure 2. Salt precipitation in the 500-nm deep network and 5-µm deep pores during CO_2 injection at constant pressure (1 MPa). (**a**–**d**) are recorded images of salt precipitation at approximately 1, 3, 4 and 12 s. The white particles are the precipitated salt contents. The yellow and red arrows indicate salt precipitations in the 500-nm deep network and 5-µm deep pores.

We used linear fit to estimate the salt participation rates of the latter two stages. In the fast salt precipitation stage (1.6 to 4.7 s), the occupancy of solid salt particles grows at a fitting rate of 9.1 ± 0.5%/s (R^2 = 0.95) (Figure 3c). In the slow salt precipitation stage (after 4.7 s), the growth rate of the occupancy of solid salt particles is fitted at 0.37 ± 0.02%/s (R^2 = 0.89) (Figure 3d). Therefore, the salt precipitation rate in the fast salt precipitation stage is ~24.6 times of the growth rate in the slow salt precipitation stage. We defined a precipitation rate ratio I to quantify this difference (i.e., $I = 24.6$).

The three stages of the salt precipitation dynamics in porous media have also been detected in a recent microfluidic study to identify CO_2-injection-caused salt precipitation in deep saline aquifers [34] The pore size in the literature is defined by the height of the micropillars to be 25 µm. The salt growth rate in the fast salt precipitation stage is measured to be ~3.1 times that of the slow salt precipitation stage (i.e., precipitation rate ratio $I = 3.3$). The precipitation rate ratio in the microporous media is thus ~7.9 times lower than that measured in our work. A potential reason for the large discrepancy between the two cases is that the water evaporation rate at the nanoscale is much lower than that at the microscale. The previous study indicates that the liquid evaporation rate (v_g) in porous media can be calculated by [33]:

$$v_g = \frac{P_s - P_v}{\frac{R_{kn}R_{vis}}{R_{kn}+R_{vis}} + R_{if}} \quad (1)$$

where P_s is the saturation pressure of the liquid in porous media, P_v is the vapor partial pressure in gas, R_{kn} is the Knudsen flow resistance defined by the Knudsen diffusion [35], R_{vis} is the viscous flow resistance defined by the Darcy's law, R_{if} is the liquid-vapor interfacial resistance defined by the Hertz-Knudsen equation [36]. R_{kn}, R_{vis} and R_{if} for the porous media consisting of slit pores are:

$$R_{kn} = \frac{3\rho_v^{mol}\tau L_v}{4h\varphi}\sqrt{2\pi RTM} \tag{2}$$

$$R_{vis} = \frac{L_v \mu_v}{k_v} \tag{3}$$

$$R_{if} = \frac{\rho_v}{a}\sqrt{\frac{2\pi RT}{M}} \tag{4}$$

where ρ_g^{mol} is the molar density of vapor phase, τ is the tortuosity of the porous media, L_v is the total length of the porous media, h is the height of a slit pore, φ is the porosity of the porous media, R is the gas constant, T is the temperature, M is the molar mass, μ_v is the vapor viscosity, k_v is the permeability of the vapor in porous media, ρ_v is the density of the vapor, a is a constant and in most cases equals 1. When the length of the porous media is orders of magnitude larger than the pore size (e.g., the height of a slit pore), R_{if} becomes negligible [33], and the water evaporation rate is mainly governed by R_{kn} and R_{vis}. The permeability of vapor in a porous media consisting of slit pores can be expressed as:

$$k = \frac{\varphi h^2}{12} \tag{5}$$

Based on Equations (1)–(5), the liquid evaporation rate can be further expressed as:

$$v_g = (P_s - P_v)\left(\frac{\varphi h^2}{12 L_v \mu_v} + \frac{4h\varphi}{3\tau\rho_v^{mol} L_v \sqrt{2\pi RTM}}\right) \tag{6}$$

For qualitative comparison, we considered a straight channel (i.e., $\varphi = \tau = 1$) and assumed that the water vapor partial pressure in the gas phase is low ($P_v \approx 0$) when the CO_2 pressure is high. Based on Equation (6), the minimum water evaporation rate in our case ($h = 500$ nm, $L_v = 2$ mm, $T = 343$ K) is calculated to be ~5.5 times slower than the minimum water evaporation rate in the literature ($h = 25$ μm, $L_v \approx 30$ mm, $T = 293$ K) [34]. The difference in evaporation rates (5.5 times) contributes to the difference (7.9 times) in the precipitation rate ratio between the two cases. Therefore, we deduced that in the shale nanoporous media, slow salt precipitation caused by long-term CO_2 injection is less pronounced than that in the deep saline aquifer because of a much lower water evaporation rate.

The precipitated salt from the fast salt precipitation stage contributes more significantly to the total precipitated salt in the nanoporous media than in the microporous media. In our case, we found that 23.9% volume of the 500-nm deep network is occupied by precipitated salt after ~13 s of CO_2 injection. At the fast salt precipitation stage, 20.3% of the network volume is occupied by precipitated salt particles, contributing to ~85% of the total deposited salt. In the literature [34], salt depositing at the fast salt precipitation stage contributes to ~77% of the total precipitated salt. The dynamics of the fast salt precipitation are governed by nucleation. Previous research points out that heterogeneous nucleation has a much lower nucleation energy barrier than homogenous nucleation [37]. In the porous media, the pore surface provides favorable nucleation sites, and heterogeneous nucleation plays a dominant role. As the surface-to-volume ratio scales with L, nanoporous media has a much larger surface-to-volume ratio than microporous media, providing a higher density of nucleation sites for heterogeneous nucleation. Therefore, salt prefers to precipitate early in the shale nanoporous media due to heterogeneous nucleation.

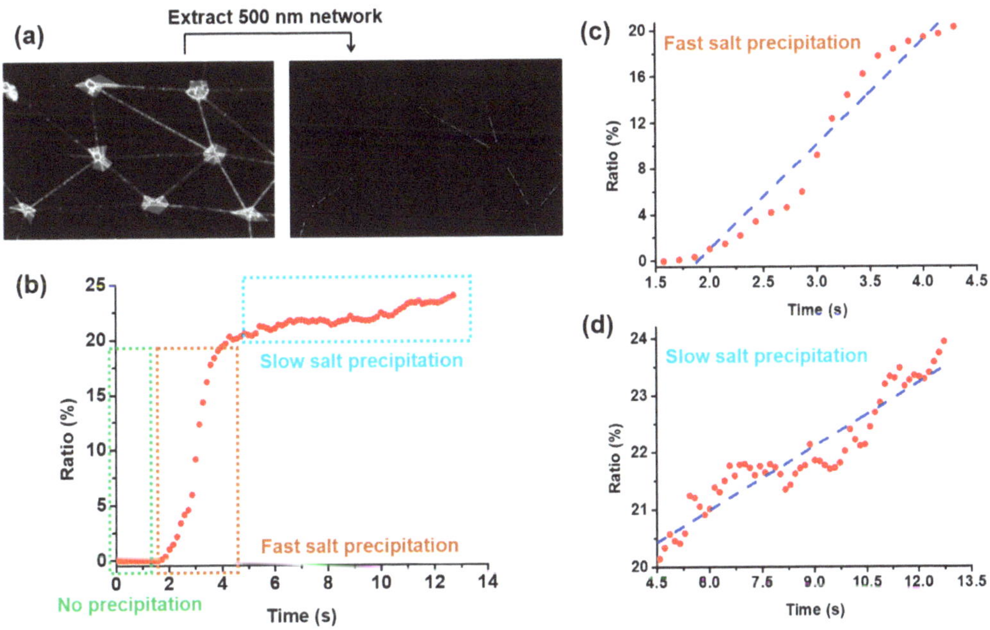

Figure 3. Salt precipitation dynamics in the 500-nm deep network. (**a**) Image analysis to extract precipitated salt particles in the 500-nm deep network. (**b**) the occupancy of the salt particles in the 500-nm deep network. The ratio is defined as the total volume of the precipitated salt divided by the total volume of the 500-nm deep network in the observation area. (**c**) Fast salt precipitation and its linear fitting result (slope = 9.1 ± 0.5, R^2 = 0.95). (**d**) Slow salt precipitation and its linear fitting result (slope = 0.37 ± 0.02, R^2 = 0.89).

During the relatively low-pressure CO_2 injection (1 MPa), we realized that the 50-nm deep network remains liquid-filled even through the gas pressure is larger than the capillary pressure. We injected high-pressure CO_2 (5 MPa) from the opposite main channel to validate the connectivity of the 50-nm deep network and test the potential of salt precipitation at the 50-nm scale. We found the salt solution in the 50-nm deep network is displaced by CO_2 due to a large pressure gradient. During the high-pressure gas injection, we still did not detect salt particles nucleating and growing in the 50-nm deep network. A potential reason is that salt nucleation is unfavorable at the 50-nm scale due to the energy barrier for forming stable nuclei.

The classical nucleation theory [38] provides qualitative insight into the experimental result. In the classical nucleation theory, the free energy change (ΔG) of the nucleus formation includes the free energy change due to the nucleus interface (ΔG_s) and the free energy change due to the nucleus volume (ΔG_v):

$$\Delta G = \Delta G_s + \Delta G_v \tag{7}$$

Here ΔG_s is positive and inhibits the nucleation, as the additional interfacial energy generated by the nucleus is unfavorable in a thermodynamic system. ΔG_v is negative and promotes the nucleation, because ions have lower chemical potential in the precipitated salts (μ_n) compared to in solution (μ_s) at supersaturation. Equation (7) can be further derived as a function of the nucleus radius (r):

$$\Delta G = 4\pi r^2 \gamma - \frac{4\pi r^3}{3V_m}(\mu_s - \mu_n) \tag{8}$$

Here γ is the interfacial tension between nucleus and solution, V_m is the volume of the unit cell in a bulk salt crystal. The schematic of the free energy of the nucleus changing with nucleus radius is shown in Figure 4. A free energy change barrier (ΔG_{max}) exists for nucleation, and the corresponding nucleus radius is the critical nucleus radius (r_c):

$$r_c = \frac{2\gamma V_m}{\mu_s - \mu_n} \tag{9}$$

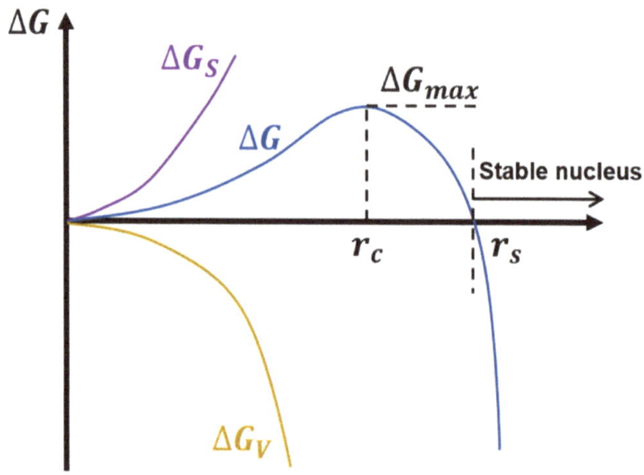

Figure 4. Schematic of the free energy change of a growing nucleus.

When the radius of the initially precipitated salt nucleus is smaller than r_c, the nucleus will not grow into a salt crystal. In addition, the free energy change needs to be negative ($\Delta G < 0$) for the growth of the nucleus to be thermodynamically favorable. Therefore, the minimum stable nucleus radius (r_s) is determined by $\Delta G = 0$:

$$r_s = \frac{3\gamma V_m}{\mu_s - \mu_n} \tag{10}$$

The critical nucleus radius changes with the composition, concentration and supersaturation conditions [39]. As we used a mixed salt solution in our experiments, determining the accurate salt nucleus size is challenging. The main composition of the simulated formation water is sodium chloride. The concentration of sodium chloride is close to saturation. Therefore, we can use the critical nucleus radius of sodium chloride as a reference. In a previous experimental study, aqueous sodium chloride was introduced to carbon nanotubes, and real-time electron microscopy at atomic resolution was used to detect the sodium chloride nucleation and precipitation [40]. The size of a stable sodium chloride nucleus is found to be on the order of several nanometers. Compared to the 500-nm deep network, the 50-nm deep network is only one order of magnitude larger than the stable sodium chloride nucleus size. The salt nucleus, once formed under perturbation, is less likely to be large and stable in the 50-nm deep network compared to the 500-nm deep network due to the confinement effect.

Another potential reason for the salt not precipitating in the 50-nm deep network during high-pressure CO_2 injection is that the salt solution was displaced by CO_2 before water evaporated to allow salt nucleation. We estimated that the minimum water evaporation rate in the 500-nm deep network is ~14.7 times faster than that in the 50-nm deep network from Equation (6). Since the CO_2 displacing salt solution happened within a similar time-lapse (~10 s) in both cases, the slow evaporation rate in the 50-nm deep network results in the salt solution being displaced by CO_2 before any nucleation could ever happen. In general, we believed that salt precipitation is unfavorable in the relatively small nanometer pores in shale matrix because the minimum stable nucleus size approaches the confinement size, and a large pressure gradient is needed to overcome the capillarity and flow resistance during CO_2 injection.

3.2. Salt Precipitation in the Micrometer Pores Connected by Nanometer Throats

During the CO_2 injection at 1 MPa, we found that salt also precipitates at the connection between the 500-nm deep network and 5-μm deep pores (Figure 5a). The precipitated salt accumulates at the inlet of the pores. We performed image analysis to evaluate the salt accumulation in three individual pores with 1, 2 and 3 inlets (Figure 5b). The three pores (Pore1, 2 and 3, Figure 5a) are ~0.35, 1 and 1.74 mm from the CO_2 injection main channel, respectively. During the CO_2 injection, the salt particles do not accumulate in the pore for the first 2.7 s, which is 1.1 s later than the initiation of salt precipitation in the 500-nm deep network. After that, the salt particles accumulate at a similar rate in all three pores, and occupy a sector area at each inlet. The fast salt precipitation in the pore ends earlier when the pore is closer to the CO_2 injection main channel (Pore1 at 5.2 s, Pore2 at 6.1 s and Pore3 at 7.7 s, Figure 5b). Meanwhile, the area that salt particles occupy per inlet is smaller when the pore is closer to the CO_2 injection main channel (Pore1 at 514 μm^2, Pore2 at 556 μm^2 and Pore3 at 638 μm^2, Figure 5b). From the experimental results, we deduced that the pore endures a larger CO_2 pressure drop when the pore is closer to the CO_2 injection main channel. The CO_2 displaces salt solution faster in pores closer to the main channel, leading to an earlier cease of salt precipitation in these pores, and thus less aggregated salt particles.

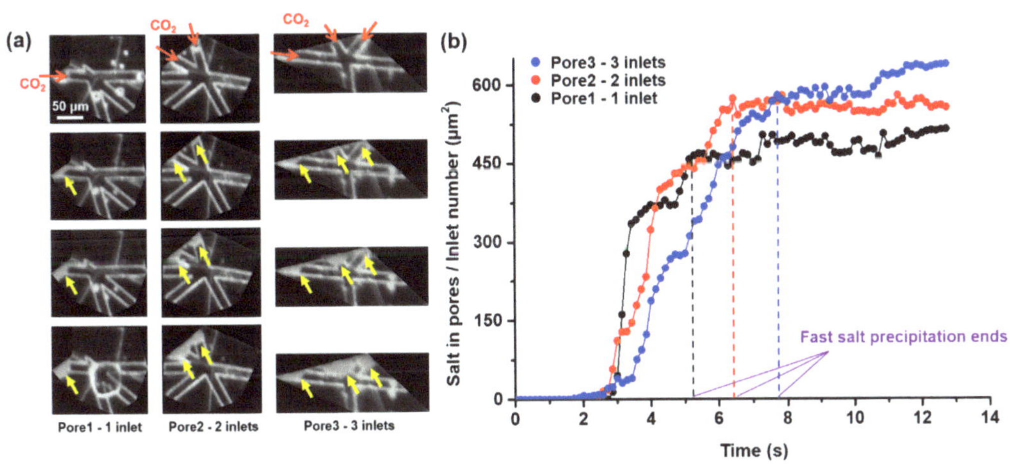

Figure 5. Salt precipitation and accumulation in the 5-μm deep pores. (**a**) Recorded salt precipitation and accumulation results in three individual pores with 1–3 inlets. (**b**) Salt particle accumulation dynamics in three pores.

Previous theoretical study of salt precipitation in porous media indicates that the capillarity drives the water flowing towards the evaporation front, leading to salt accumulation at the outlet of the porous media [11]. Similarly, in a previous experimental study, researchers found that after drying the core initially saturated with brine, precipitated salt is mainly found on the core surface [41]. In our case, when CO_2 is injected into the 5-μm deep pore through the nanochannel, the relatively large pore serves as an open space for water to evaporate and salt to accumulate.

To further analyze the salt accumulation dynamics in pores, we performed the phase-field simulation [42] to evaluate the gas chamber in the pore during CO_2 injection, as shown in Figure 6a. The 2D simulation captures the governing dimension (depth) of the channel (500 nm) and pore (5 μm). We set the pore length as 100 μm, which is the magnitude of the equivalent diameter of the pore on the chip surface. In the simulation, there is also a 500-nm outlet nanochannel for the solution to be displaced, which is not plotted in Figure 6a. We chose the fluid properties, including viscosity, density and interfacial tension of CO_2 and salt solution at 70 °C and 1 MPa, and set the CO_2 injection pressure as 1 MPa. It is noted that during the experiment, we can only observe the salt precipitation from the top side of the nanofluidic chip. The simulation applies the geometry of the governing dimension (depth). We only used simulation results for qualitative analysis instead of quantitative comparison.

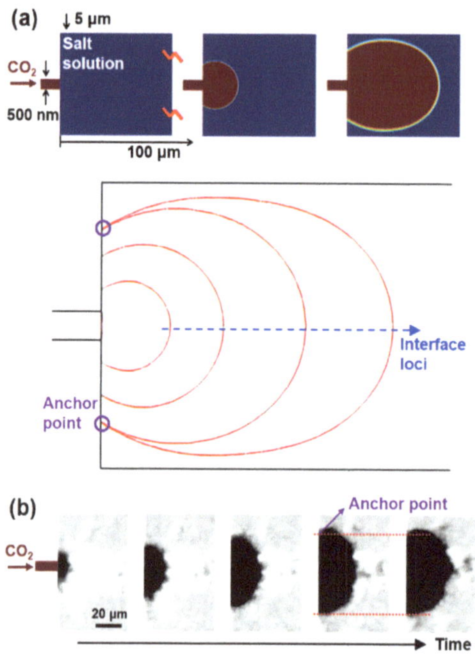

Figure 6. (**a**) Simulation of CO_2 displacing salt solution at the connection between a 500-nm channel and a 5-μm pore. The loci of the CO_2-water interface are plotted by combing multiple frames of simulation results. (**b**) Experimental results of precipitated salt accumulating at a pore inlet. The precipitated salt is colored black for clarity.

The CO_2 displacing salt solution produces a growing gas chamber, as shown in Figure 6a. At the CO_2-water interface, water evaporation causes salt precipitation and accumulation. As the CO_2 displacement front remains curved, the salt precipitation front also grows in a curved shape and accumulates in a sector area (Figures 5a and 7b). In the simulation, the effect of accumulated salt particles on the CO_2 displacing salt solution is not considered. Therefore, the center of the CO_2-water interface moves the fastest during the CO_2 injection (Figure 6a), leading to a center-stretched arc profile. However, in the experiment, we found that the salt accumulation rate at the center is lower than what we expected from the simulation (Figure 6b). The salt accumulation front is actually a center-compressed arc profile. Therefore, the accumulated salt particles at the pore inlet impede CO_2 flow rate severely. In addition, the simulation results indicate that an anchor point of the interface exists during the CO_2 displacing salt solution (Figure 6a). In the experiment, the anchor point is also identified during the salt accumulation, further proving that salt precipitation in pores mainly happens at the CO_2-solution interface with CO_2 injection.

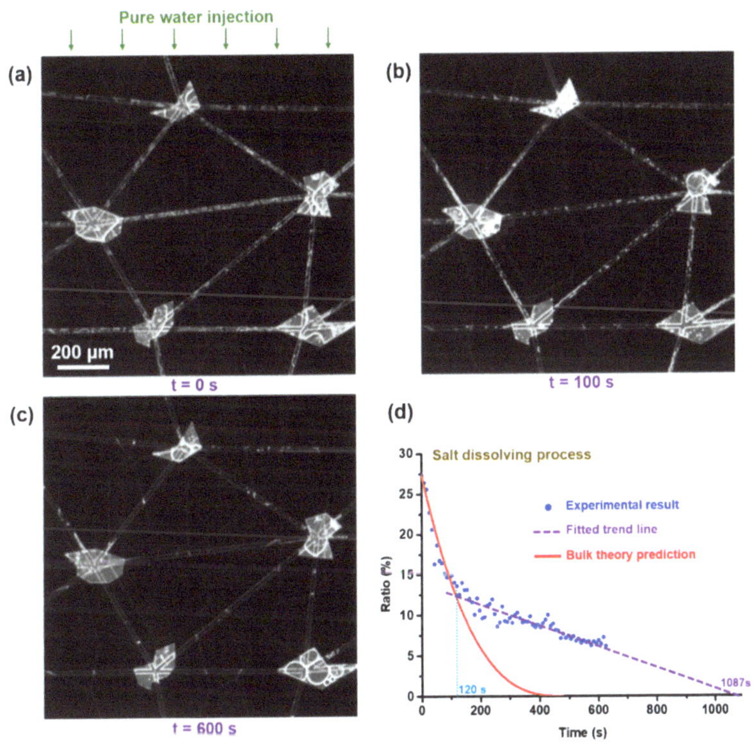

Figure 7. Salt dissolution dynamics in the 500-nm deep network. (**a**–**c**) are recorded images of salt dissolution at 0, 100 and 600 s. (**d**) Experimental results of the salt dissolution process, classical bulk dissolution theory prediction and fitted trend line based on experimental results.

3.3. *Salt Dissolution in the Nanoporous Media*

We injected pure water into the salt-contaminated chip to study the salt dissolution dynamics in the nanoporous media (Figure 7). The water injection is at constant pressure (1 MPa), and water slowly dissolves salt particles in the 500-nm deep network. The salt precipitation process takes ~13 s during the CO_2 injection, but salt dissolution needs more than 600 s at the same injection pressure (1 MPa). After injecting water for 625.7 s, the occupancy of the solid salt particles in the 500-nm deep network reduces from an initial 27.3% to 6.2%. In the experiment, we did not record long enough to capture the moment when all salt particles get dissolved because of the limited camera memory. We estimated the time needed for salt dissolution in the 500-nm deep network to be ~1087 s by linearly fitting the slow dissolution stage (Figure 7d). The duration of salt dissolution is 83.6 times longer than the duration of salt precipitation. At 70 °C and 1 MPa, the water viscosity is 23.5 times larger than the CO_2 viscosity. Therefore, the water flow rate in the 500-nm deep network is much lower than the CO_2 flow rate. The difference in flow rate contributes to the discrepancy in the time required for salt precipitation and dissolution. The remaining discrepancy (~3.6 times) should be caused by different governing mechanisms of salt precipitation (governed by salt nucleation and water evaporation rate) and dissolution (governed by ion hydration and diffusion).

The salt dissolution dynamics at bulk have been studied early [43], and is characterized by the equation below:

$$V\frac{dw}{dt} = -Kw^{2/3}(w_s - w_0 + w) \tag{11}$$

where V is the volume of solution, w is the weight of salt crystal at time t, K is a positive constant relating to dissolution rate, w_s is the weight of salt needed to saturate liquid phase under given conditions, w_0 is the weight of salt crystal at the beginning. In our case, the pure water is continuously injected. We can assume the salt concentration in the solution is small, and its change is negligible. The Equation (11) can be further simplified as [43]:

$$V\frac{dw}{dt} = -K_1 w^{2/3} \tag{12}$$

where K_1 is also a positive constant relating to dissolution rate, and is different from K. The integration of Equation (12) with the initial condition ($t = 0$, $w = w_0$) gives the expression of salt crystal weight changing with time:

$$w = \left(w_0^{1/3} - \frac{K_1 t}{3}\right)^3 \tag{13}$$

In our experiment, the weight of salt crystal is equivalent as the occupancy of the solid salt particles in the 500-nm deep network. Based on Equation (13), the salt dissolution dynamics is described as:

$$R = \left(R_0^{1/3} - K_2 t\right)^3 \tag{14}$$

where K_2 is also a positive constant relating to dissolution rate, and is different from K and K_1. By fitting the K_2 with our experimental results, we found the classical dissolution theory at bulk only predicts the early fast dissolution stage ($t \leq 120$ s) in the nanoporous media, as shown in Figure 7d. The dissolution process in the 500-nm deep network is expected to take twice as long as the duration predicted by the classical theory (~480 s). Therefore, in the nanoporous media, classical bulk dissolution theory could considerably overestimate the actual dissolution rate.

The lower diffusion coefficient of hydrated ions in the nanoporous media potentially causes the slow dissolution stage ($t > 120$ s) found in our experiment. Previous theoretical studies often indicate that molecular diffusivity in confinement varies from its bulk value when the confinement size reduces to serval times of the molecular diameter (i.e., sub-10 nm) [44,45]. Experimental studies find that even at 10^2 nm confinement size, the molecular diffusivity can be reduced significantly. For example, researchers used the conical nanochannel with an orifice radius at 872 nm to measure the diffusivity of the FITC fluorescent dye molecule (molecular diameter at ~2 nm) in confinement. They found that the diffusivity of the FITC molecule reduces from 3.32×10^{-7} cm^2/s in bulk solution to a minimum of 4.61×10^{-8} cm^2/s in confinement [46]. Here the diameters of the hydrated sodium ion and the hydrated chloride ion are ~0.7 nm [47]. The inference is that, for the salt dissolution in the 500-nm network, the initial fast dissolution ($t \leq 120$ s) is governed by the hydration of ions, while the later slow dissolution is governed by the diffusion of hydrated ions at the 500-nm scale. Overall, our experimental findings prove that classical dissolution theory fails to predict the salt dissolution dynamics at the nanometer pore scale. Further theoretical study is needed to model the entire dissolution process in the nanoporous media.

4. Conclusions

In this work, we developed a multiscale shale micromodel (50 nm, 500 nm and 5 μm) to characterize the salt precipitation during CO_2 injection, as well as the salt dissolution during water injection. For the salt precipitation in the nanoporous media, we found that salt precipitation mainly happens in the 500-nm deep network. Two stages of salt precipitation (fast and slow) are observed. The precipitation rate ratio of the two stages is ~7.9 times larger than the ratio measured in the microporous media. The slow water evaporation and fast heterogeneous nucleation in nanoporous media potentially contribute to the difference in the precipitation rate ratio in nanoporous and microporous media.

In addition to the nanometer pore network, salt precipitation also happens at the connection between the 500-nm deep network and 5-μm deep pores. Salt particles accumulate in a sector region that aligns with the CO_2 displacement front. During the water injection, the salt dissolution in the 500-nm network takes approximately twice as long as the prediction from the classical bulk dissolution theory. The theory only predicts the initial fast dissolution in the experiment, but misses the later slow dissolution stage, indicating the failure of bulk theory at the nanoscale.

Overall, this work provides nanometer pore-scale insight into salt precipitation and dissolution during CO_2 and water injection. The experimental findings shed light on CO_2 injection for shale oil recovery when high salinity water exists. We conducted preliminary theoretical analyses of experimental results. Further theoretical modeling work to dig into the fundamental mechanisms of salt precipitation and dissolution dynamics at the nanoscale is urgently needed. In addition, the injected CO_2 is often at the supercritical condition in shale reservoirs. Characterizing the effect of supercritical CO_2 on salt precipitation is also required.

Author Contributions: Software, H.D.; Investigation, X.H.; Resources, X.H., H.X. and L.Z. (Lei Zhang); Data curation, L.Z. (Liang Zhang) and W.S.; Writing—review & editing, H.S.; Visualization, Z.W.; Supervision, J.Y. and J.Z.; Project administration, J.Z.; Funding acquisition, J.W. and J.Z. All authors have read and agreed to the published version of the manuscript.

Funding: The authors gratefully acknowledge the generous support from the National Natural Science Foundation of China (No. 52174051), Excellent Young Scholars of Shandong Province (No. 2022HWYQ-072), Guanghua Scholars of China University of Petroleum (East China) (No. 20210002) as well as research fund provided by the Exploration and Development Research Institute Sinopec Jianghan Oilfield Company.

Conflicts of Interest: The authors declare no conflict of interest.

References

1. Jia, B.; Tsau, J.-S.; Barati, R. A review of the current progress of CO_2 injection EOR and carbon storage in shale oil reservoirs. *Fuel* **2019**, *236*, 404–427.
2. Mukhina, E.; Cheremisin, A.; Khakimova, L.; Garipova, A.; Dvoretskaya, E.; Zvada, M.; Kalacheva, D.; Prochukhan, K.; Kasyanenko, A.; Cheremisin, A. Enhanced Oil Recovery Method Selection for Shale Oil Based on Numerical Simulations. *ACS Omega* **2021**, *6*, 23731–23741. [CrossRef] [PubMed]
3. Burrows, L.C.; Haeri, F.; Cvetic, P.; Sanguinito, S.; Shi, F.; Tapriyal, D.; Goodman, A.; Enick, R.M. A Literature Review of CO_2, Natural Gas, and Water-Based Fluids for Enhanced Oil Recovery in Unconventional Reservoirs. *Energy Fuels* **2020**, *34*, 5331–5380. [CrossRef]
4. Feng, Q.; Xu, S.; Xing, X.; Zhang, W.; Wang, S. Advances and challenges in shale oil development: A critical review. *Adv. Geo-Energy Res.* **2020**, *4*, 406–418. [CrossRef]
5. Wu, W.; Zoback, M.D.; Kohli, A.H. The impacts of effective stress and CO_2 sorption on the matrix permeability of shale reservoir rocks. *Fuel* **2017**, *203*, 179–186. [CrossRef]
6. Alfarge, D.; Wei, M.; Bai, B. Factors Affecting CO_2-EOR in Shale-Oil Reservoirs: Numerical Simulation Study and Pilot Tests. *Energy Fuels* **2017**, *31*, 8462–8480. [CrossRef]
7. Yu, H.; Xu, H.; Fu, W.; Lu, X.; Chen, Z.; Qi, S.; Wang, Y.; Yang, W.; Lu, J. Extraction of shale oil with supercritical CO_2: Effects of number of fractures and injection pressure. *Fuel* **2021**, *285*, 118977. [CrossRef]
8. Khan, H.J.; Spielman-Sun, E.; Jew, A.D.; Bargar, J.; Kovscek, A.; Druhan, J.L. A Critical Review of the Physicochemical Impacts of Water Chemistry on Shale in Hydraulic Fracturing Systems. *Environ. Sci. Technol.* **2021**, *55*, 1377–1394. [CrossRef]
9. Zhang, Y.; Ju, B.; Zhang, M.; Wang, C.; Zeng, F.; Hu, R.; Yang, L. The effect of salt precipitation on the petrophysical properties and the adsorption capacity of shale matrix based on the porous structure reconstruction. *Fuel* **2022**, *310*, 122287.
10. Fakher, S.; Imqam, A. Asphaltene precipitation and deposition during CO_2 injection in nano shale pore structure and its impact on oil recovery. *Fuel* **2019**, *237*, 1029–1039. [CrossRef]
11. Miri, R.; Hellevang, H. Salt precipitation during CO_2 storage—A review. *Int. J. Greenh. Gas Control* **2016**, *51*, 136–147. [CrossRef]
12. Kumar, S.; Foroozesh, J.; Edlmann, K.; Rezk, M.G.; Lim, C.Y. A comprehensive review of value-added CO_2 sequestration in subsurface saline aquifers. *J. Nat. Gas Sci. Eng.* **2020**, *81*, 103437. [CrossRef]
13. Lebedev, M.; Zhang, Y.; Sarmadivaleh, M.; Barifcani, A.; Al-Khdheeawi, E.; Iglauer, S. Carbon geosequestration in limestone: Pore-scale dissolution and geomechanical weakening. *Int. J. Greenh. Gas Control* **2017**, *66*, 106–119. [CrossRef]
14. De Silva, P.N.K.; Ranjith, P.G. A study of methodologies for CO_2 storage capacity estimation of saline aquifers. *Fuel* **2012**, *93*, 13–27. [CrossRef]

15. Peysson, Y.; André, L.; Azaroual, M. Well injectivity during CO_2 storage operations in deep saline aquifers—Part 1: Experimental investigation of drying effects, salt precipitation and capillary forces. *Int. J. Greenh. Gas Control* **2014**, *22*, 291–300. [CrossRef]
16. Ott, H.; Snippe, J.; de Kloe, K. Salt precipitation due to supercritical gas injection: II. Capillary transport in multi porosity rocks. *Int. J. Greenh. Gas Control* **2021**, *105*, 103233. [CrossRef]
17. Oh, J.; Kim, K.-Y.; Han, W.S.; Kim, T.; Kim, J.-C.; Park, E. Experimental and numerical study on supercritical CO_2/brine transport in a fractured rock: Implications of mass transfer, capillary pressure and storage capacity. *Adv. Water Resour.* **2013**, *62*, 442–453. [CrossRef]
18. Kim, K.-Y.; Han, W.S.; Oh, J.; Kim, T.; Kim, J.-C. Characteristics of Salt-Precipitation and the Associated Pressure Build-Up during CO_2 Storage in Saline Aquifers. *Transp. Porous Media* **2012**, *92*, 397–418. [CrossRef]
19. Alhamad, L.; Miskimins, J. Minimizing calcium lactate precipitation via the addition of gluconate ions for matrix acidizing with lactic acid. *J. Pet. Sci. Eng.* **2022**, *218*, 110995. [CrossRef]
20. Wang, H.; Liu, W. Research on numerical simulation method of salt dissolution and recrystallization of inter-salt shale oil reservoir. *J. Pet. Sci. Eng.* **2022**, *213*, 110464. [CrossRef]
21. Zhang, D.; Kang, Y.; Selvadurai, A.P.S.; You, L. Experimental Investigation of the Effect of Salt Precipitation on the Physical and Mechanical Properties of a Tight Sandstone. *Rock Mech. Rock Eng.* **2020**, *53*, 4367–4380. [CrossRef]
22. Zhang, Y.; Zhang, M.; Mei, H.; Zeng, F. Study on salt precipitation induced by formation brine flow and its effect on a high-salinity tight gas reservoir. *J. Pet. Sci. Eng.* **2019**, *183*, 106384. [CrossRef]
23. Shao, J.; You, L.; Kang, Y.; Chen, M. Experimental investigation of effect of salt crystallization on the mechanical strength of shale. *J. Pet. Sci. Eng.* **2022**, *213*, 110366. [CrossRef]
24. Kim, M.; Sell, A.; Sinton, D. Aquifer-on-a-Chip: Understanding pore-scale salt precipitation dynamics during CO_2 sequestration. *Lab Chip* **2013**, *13*, 2508–2518. [CrossRef] [PubMed]
25. He, D.; Jiang, P.; Xu, R. Pore-Scale Experimental Investigation of the Effect of Supercritical CO_2 Injection Rate and Surface Wettability on Salt Precipitation. *Environ. Sci. Technol.* **2019**, *53*, 14744–14751. [CrossRef]
26. Zhong, J.; Riordon, J.; Zandavi, S.H.; Xu, Y.; Persad, A.H.; Mostowfi, F.; Sinton, D. Capillary Condensation in 8 nm Deep Channels. *J. Phys. Chem. Lett.* **2018**, *9*, 497–503. [CrossRef]
27. Yang, Q.; Jin, B.; Banerjee, D.; Nasrabadi, H. Direct visualization and molecular simulation of dewpoint pressure of a confined fluid in sub-10 nm slit pores. *Fuel* **2019**, *235*, 1216–1223. [CrossRef]
28. Zhong, J.; Alibakhshi, M.A.; Xie, Q.; Riordon, J.; Xu, Y.; Duan, C.; Sinton, D. Exploring Anomalous Fluid Behavior at the Nanoscale: Direct Visualization and Quantification via Nanofluidic Devices. *Acc. Chem. Res.* **2020**, *53*, 347–357. [CrossRef]
29. Xie, Q.; Alibakhshi, M.A.; Jiao, S.; Xu, Z.; Hempel, M.; Kong, J.; Park, H.G.; Duan, C. Fast water transport in graphene nanofluidic channels. *Nat. Nanotechnol.* **2018**, *13*, 238–245. [CrossRef]
30. Wang, L.; Boutilier, M.S.; Kidambi, P.R.; Jang, D.; Hadjiconstantinou, N.G.; Karnik, R. Fundamental transport mechanisms, fabrication and potential applications of nanoporous atomically thin membranes. *Nat. Nanotechnol.* **2017**, *12*, 509. [CrossRef]
31. Zhong, J.; Abedini, A.; Xu, L.; Xu, Y.; Qi, Z.; Mostowfi, F.; Sinton, D. Nanomodel visualization of fluid injections in tight formations. *Nanoscale* **2018**, *10*, 21994–22002. [CrossRef] [PubMed]
32. Zhong, J.; Zhao, Y.; Lu, C.; Xu, Y.; Jin, Z.; Mostowfi, F.; Sinton, D. Nanoscale Phase Measurement for the Shale Challenge: Multicomponent Fluids in Multiscale Volumes. *Langmuir* **2018**, *34*, 9927–9935. [CrossRef] [PubMed]
33. Jatukaran, A.; Zhong, J.; Persad, A.H.; Xu, Y.; Mostowfi, F.; Sinton, D. Direct Visualization of Evaporation in a Two-Dimensional Nanoporous Model for Unconventional Natural Gas. *ACS Appl. Nano Mater.* **2018**, *1*, 1332–1338. [CrossRef]
34. Ho, T.-H.M.; Tsai, P.A. Microfluidic salt precipitation: Implications for geological CO_2 storage. *Lab Chip* **2020**, *20*, 3806–3814. [CrossRef] [PubMed]
35. Gruener, S.; Huber, P. Knudsen Diffusion in Silicon Nanochannels. *Phys. Rev. Lett.* **2008**, *100*, 064502. [CrossRef] [PubMed]
36. Persad, A.H.; Ward, C.A. Expressions for the Evaporation and Condensation Coefficients in the Hertz-Knudsen Relation. *Chem. Rev.* **2016**, *116*, 7727–7767. [CrossRef]
37. Khaleghi, A.; Sadrameli, S.M.; Manteghian, M. Thermodynamic and kinetics investigation of homogeneous and heterogeneous nucleation. *Rev. Inorg. Chem.* **2020**, *40*, 167–192. [CrossRef]
38. Gebauer, D.; Raiteri, P.; Gale, J.D.; Cölfen, H. On classical and non-classical views on nucleation. *Am. J. Sci.* **2018**, *318*, 969–988. [CrossRef]
39. Zimmermann, N.E.R.; Vorselaars, B.; Quigley, D.; Peters, B. Nucleation of NaCl from Aqueous Solution: Critical Sizes, Ion-Attachment Kinetics, and Rates. *J. Am. Chem. Soc.* **2015**, *137*, 13352–13361. [CrossRef]
40. Nakamuro, T.; Sakakibara, M.; Nada, H.; Harano, K.; Nakamura, E. Capturing the Moment of Emergence of Crystal Nucleus from Disorder. *J. Am. Chem. Soc.* **2021**, *143*, 1763–1767. [CrossRef]
41. Desarnaud, J.; Derluyn, H.; Molari, L.; de Miranda, S.; Cnudde, V.; Shahidzadeh, N. Drying of salt contaminated porous media: Effect of primary and secondary nucleation. *J. Appl. Phys.* **2015**, *118*, 114901. [CrossRef]
42. Guangpu, Z.; Aifen, L. Interfacial dynamics with soluble surfactants: A phase-field two-phase flow model with variable densities. *Adv. Geo-Energy Res.* **2020**, *4*, 86–98.
43. Noyes, A.A.; Whitney, W.R. The rate of solution of solid substances in their own solutions. *J. Am. Chem. Soc.* **1897**, *19*, 930–934. [CrossRef]

44. Wang, H.; Wang, X.; Jin, X.; Cao, D. Molecular Dynamics Simulation of Diffusion of Shale Oils in Montmorillonite. *J. Phys. Chem. C* **2016**, *120*, 8986–8991. [CrossRef]
45. Zhang, W.; Feng, Q.; Wang, S.; Xing, X. Oil diffusion in shale nanopores: Insight of molecular dynamics simulation. *J. Mol. Liq.* **2019**, *290*, 111183. [CrossRef]
46. Zhang, L.-X.; Cao, X.-H.; Cai, W.-P.; Li, Y.-Q. Observations of the Effect of Confined Space on Fluorescence and Diffusion Properties of Molecules in Single Conical Nanopore Channels. *J. Fluoresc.* **2011**, *21*, 1865–1870. [CrossRef]
47. Tansel, B.; Sager, J.; Rector, T.; Garland, J.; Strayer, R.F.; Levine, L.; Roberts, M.; Hummerick, M.; Bauer, J. Significance of hydrated radius and hydration shells on ionic permeability during nanofiltration in dead end and cross flow modes. *Sep. Purif. Technol.* **2006**, *51*, 40–47. [CrossRef]

energies

MDPI

Article

A Practical Model for Gas–Water Two-Phase Flow and Fracture Parameter Estimation in Shale

Pin Jia [1,2,*], Langyu Niu [1,2], Yang Li [1,2] and Haoran Feng [1,2]

1 State Key Laboratory of Petroleum Resources and Prospecting, China University of Petroleum (Beijing), Beijing 102249, China
2 College of Petroleum Engineering, China University of Petroleum (Beijing), Beijing 102249, China
* Correspondence: jiapin1990@163.com

Abstract: The gas flow in shale reservoirs is controlled by gas desorption diffusion and multiple flow mechanisms in the shale matrix. The treatment of hydraulic fracturing injects a large amount of fracturing fluids into shale reservoirs, and the fracturing fluids can only be recovered by 30~70%. The remaining fracturing fluid invades the reservoir in the form of a water invasion layer. In this paper, by introducing the concept of a water invasion layer, the hydraulic fracture network is di-vided into three zones: major fracture, water invasion layer and stimulated reservoir volume (SRV). The mathematical model considering gas desorption, the water invasion layer and gas–water two-phase flow in a major fracture is established in the Laplace domain, and the semi-analytical solution method is developed. The new model is validated by a commercial simulator. A field case from WY shale gas reservoir in southwestern China is used to verify the utility of the model. Several key parameters of major fracture and SRV are interpreted. The gas–water two-phase flow model established in this paper provides theoretical guidance for fracturing effectiveness evaluation and an efficient development strategy of shale gas reservoirs.

Keywords: shale gas; gas–water two-phase; fluid model; water invasion layer; parameter estimation of fracture network

1. Introduction

Due to the large amount of fracturing fluids injected, the hydraulic fracture networks and the surrounding matrix in the shale reservoir are complex. After a shut-in period, the fracturing fluids will create a water blockage in the shale matrix near fractures, which essentially is formation damage with a lower permeability. In this paper, the fracture network and matrix in shale gas reservoirs are classified into three systems: stimulated reservoir volume (SRV), water invasion layer and hydraulic fracture. The current gas–water two-phase flow model does not consider the influence of the water invasion layer on gas well production. Thus, it is difficult to accurately interpret the parameters of a complex fracture network and evaluate the fracturing effectiveness.

In 1998, Wattenbarger et al. [1] proposed a linear flow model for production dynamic analysis of tight gas wells. In 2010, Bello and Wattenbarger [2] developed a cylindrical dual-porosity model, and this model proved to be useful for productivity analysis of fractured horizontal wells, and the flow regimes during production were analyzed using this model. In 2011, Al-Ahmadi and Wattenbarger [3] considered secondary fractures of the reservoir based on the dual-porosity model, and the SRV was therefore considered as a tri-porosity model. In the model, the quasi-steady and unsteady flow transfer between matrix-secondary fracture and primary fracture are considered. The analytical solution of the model is obtained in Laplace space, and the flow regimes were obtained. The model shows the strong ability to interpret the complex fracture network in shale reservoirs. In 2011, Brown et al. [4] established a model with the assumption that the fluid from the outer zone was supplied to the inner zone, and then flowed into the main fracture from the inner zone, and finally into the wellbore. A classical trilinear flow model was set up, and several local approximate solutions of the model were obtained, which were suitable for linear and bilinear flow stages. At present, this model has been widely used in the production performance analysis of unconventional oil and gas wells. Then, Stalgorova and Mattar [5] divided the fractured horizontal well model into five regions by further considering the unstimulated regions between the various levels of fracture and gave the solution in the Laplace domain. Ai [6] established a prediction model for shale gas well productivity considering the contribution of unstimulated reservoir volume (unSRV) to the productivity. From the literatures, the current studies rarely consider the influence of water invasion layer in the modeling of fractured horizontal wells and the estimation of the parameters of fracture network in shale reservoirs. AlQuaimi [7] proposed a new capillary-number definition for fractures that incorporates geometrical characterization of the fracture, dependent on the force balance on a trapped ganglion.

In this paper, the SRV, water invasion layer and hydraulic fracture are finely divided. The influence of the water invasion layer on production is considered in reservoir modeling. A new model considering gas adsorption of shale matrix, water invasion layer and gas–water two-phase flow in hydraulic fractures is established and solved semi-analytically in this paper. Additionally, a fitting method of production data is proposed for the estimation of fracture network parameters of shale reservoirs.

2. Mathematical Model for Gas-Water Two-Phase Flow

2.1. Physical Model and Assumptions

In Figure 1, the white zone is the hydraulic fracture, the blue zone is the water invasion layer and the dark gray zone is the SRV, also known as the inner zone. The light gray zone is the unSRV, also known as the outer zone. Since we focus on the gas–water two-phase flow period at the early stage of gas well production, the linear flow in the unSRV (outer zone) has little influence on the production. The flow in the unSRV is not considered, as shown in Figure 1.

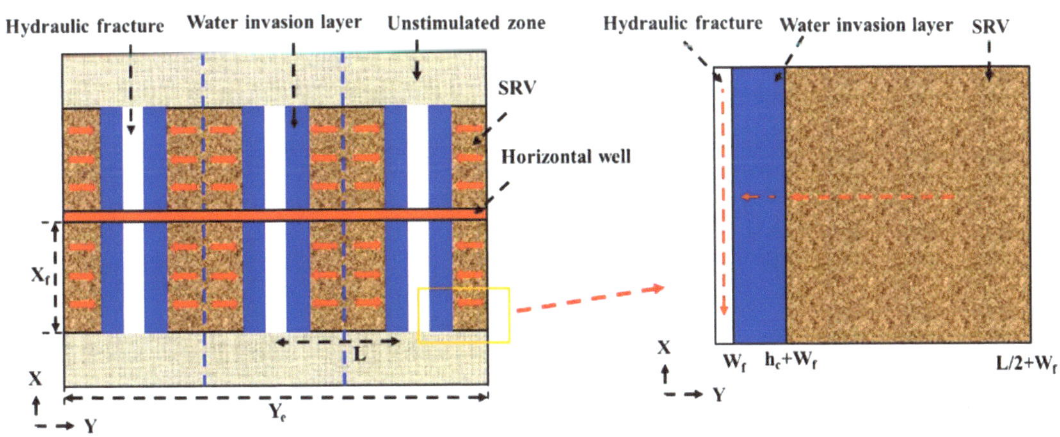

Figure 1. The schematic diagram of physical model.

With the production, the gas in the SRV flows linearly from the matrix to the water invasion layer, then reaches the hydraulic fracture, and finally flows into the wellbore. The following assumptions are made:

(1) The reservoir is of equal thickness, homogeneous and closed. The horizontal well maintains constant pressure production.

(2) Gas flow is considered to be single-phase in the SRV, and gas–water two-phase flow is considered in the water invasion layer and hydraulic fracture.
(3) The effects of gas desorption and slippage are considered in the shale matrix, and the stress sensitivity of hydraulic fracture is considered.

2.2. Single-Phase Gas Flow in Stimulated Reservoir

As shown in Figure 2, both free gas and adsorbed gas exist in the shale matrix. As the reservoir pressure drops, the free gas and adsorbed gas in the matrix expand to provide gas.

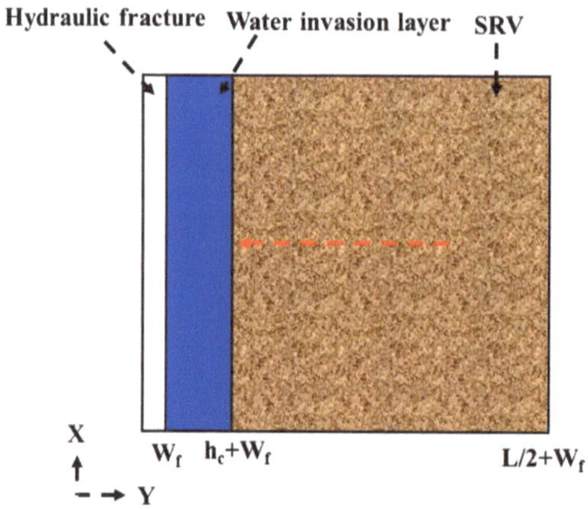

Figure 2. Single-phase gas flow in the SRV.

The single-phase gas flow equation can be written as follows:

$$\frac{\partial}{\partial y}(-\rho_{sc}v_m) = 0.0864[\frac{\partial(\rho\phi_m)}{\partial t} + \frac{\partial V}{\partial t}] \quad (1)$$

where, ρ_{sc} is the gas density in standard condition, kg/m^3; ϕ_m is matrix porosity, dimensionless; t is time, d; v_m is the gas flow velocity, m/d; and V is the adsorption capacity of the SRV, m^3. The second term on the right side of the equation represents the mass of gas desorbed from the surface of shale matrix per unit volume per unit time.

According to the Langmuir equation of gas adsorption and desorption [8], the amount of desorption at any point in the formation is closely related to the matrix pressure at that point. Thus, the gas isothermal adsorption equation is introduced as follows:

$$V = \rho_{sc}\frac{V_L p_m}{p_L + p_m} \quad (2)$$

where, p_m is the matrix pressure, MPa; V is adsorption capacity, m^3; V_L is the Langmuir volume, m^3/m^3; p_L is Langmuir pressure, MPa; and ρ_{sc} is the gas density in standard condition, kg/m^3.

By introducing the pseudo-pressure:

$$\psi_m(p_m) = 2\int_{p_i}^{p_m} \frac{p}{\mu_g z}dp \quad (3)$$

The gas flow equation of the SRV is expressed as:

$$\frac{\partial^2\psi_m}{\partial y^2} = 0.0864\frac{\phi_m\mu_g(p_m)c_{tm}(p_m)}{k_m}\frac{\partial\psi_m}{\partial t} \quad (4)$$

Further, for the nonlinear term on the right side of the equation, pseudo-time is used, which is defined as:

$$t_a = \int_0^t \frac{\mu_{gi} c_{tmi}}{\mu_g(\overline{p}_m) c_{tm}(\overline{p}_m)} dt \tag{5}$$

Thus, the single-phase gas flow equation in the SRV is rearranged by using the pseudo-pressure and pseudo-time, and the gas flow model in the SRV can be obtained as follows:

$$\begin{cases} \frac{\partial^2 \psi_m}{\partial y^2} = \frac{\mu_{gi}(\phi c_t)_{mi}}{0.0864 k_m} \frac{\partial \psi_m}{\partial t_a} \\ \psi_m(y,t)|_{t_a=0} = \psi_i \\ \left. \frac{\partial \psi_m(y,t_a)}{\partial y} \right|_{y=L_F/2} = 0 \end{cases} \tag{6}$$

The inner boundary of the SRV is coupled with the water invasion layer system, and the pressure and flow rate on the inner boundary and the water invasion layer boundary should be equal, respectively:

$$\psi_m|_{y=h_c+w_F/2} = \psi_c|_{y=h_c+w_F/2} \tag{7}$$

2.3. *Gas-Water Two-Phase Flow in Water Invasion Layer*

After hydraulic fracturing, the water (fracturing fluid) invade into the shale matrix through mass transfer from the hydraulic fracture. As shown in Figure 3, This fluid is movable water during the flowback period, and gas–water two-phase flow occurs in this zone.

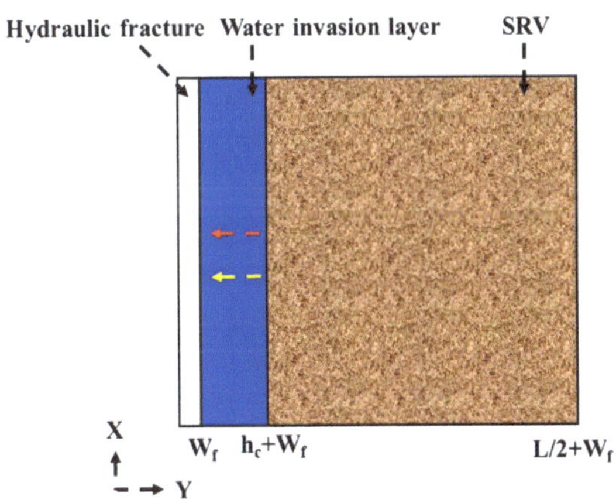

Figure 3. Gas–water two-phase fluid in water invasion layer.

The flow equation of gas phase in the water invasion layer is similar to the flow equation of gas phase in the SRV. The gas flow equation also uses pseudo-pressure and pseudo-time, and can be obtained as follows:

$$\begin{cases} \frac{\partial^2 \psi_c}{\partial y^2} = \frac{\mu_{gi}(\phi_c c_t)_{ci}}{0.0864 k_c \hat{k}_{crg}} \frac{\partial \psi_c}{\partial t_a} \\ \psi_c(y,t_a)|_{t_a=0} = \psi_i \end{cases} \tag{8}$$

where, $\hat{k}_{crg}$ is the relative permeability of the gas phase in the water invasion layer, the relative permeability is a function of gas saturation. In this paper, time is discretized into several time steps, and in each time step, an average saturation is taken, which can be treated as a constant in a single time step. ϕ_c is the porosity of water invasion layer, dimensionless; μ_g is gas viscosity, mPa·s; ψ_c is the pseudo-pressure of water invasion layer, $(\mathrm{MPa})^2/(\mathrm{mPa, s})$; and c_{tc} is the total compressibility of water invasion layer system, MPa^{-1}.

The outer boundary of the water invasion layer is coupled with the system flow model of the SRV, and the coupling condition is as follows:

$$k_m \frac{\partial \psi_m}{\partial y}\bigg|_{y=h_c+w_F/2} = k_c \frac{\partial \psi_c}{\partial y}\bigg|_{y=h_c+w_F/2} \tag{9}$$

The inner boundary is coupled to the boundary of the hydraulic fracture, so the pressure at the interface of these two zones should be equal, i.e.,

$$\psi_c(y, t_a)|_{y=w_F/2} = \psi_F \tag{10}$$

The water flow equation can be obtained as follows:

$$\begin{cases} \frac{\partial^2 p_c}{\partial y^2} = \frac{1}{0.0864} \frac{\phi_c c_{tc}^* \mu_w}{k_c \hat{k}_{crw}} \frac{\partial p_c}{\partial t} \\ p_c(y,t)|_{t=0} = p_i \\ \frac{\partial p_c}{\partial y}\Big|_{y=h_c+w_F/2} = 0 \end{cases} \tag{11}$$

Due to the coupling between the inner boundary of the water invasion layer and the hydraulic fracture system, the pressure at the interface of these two zones should be equal, i.e.,

$$p_c(y,t)|_{y=w_F/2} = p_F \tag{12}$$

2.4. Gas-Water Two-Phase Flow in Hydraulic Fracture

Similar to the water invasion layer, the hydraulic fracture is filled with fracturing fluid after fracturing stimulation, and gas–water two-phase flow occurs during flowback. As shown in Figure 4, The gas and water continuously supply the hydraulic fracture through the water invasion layer.

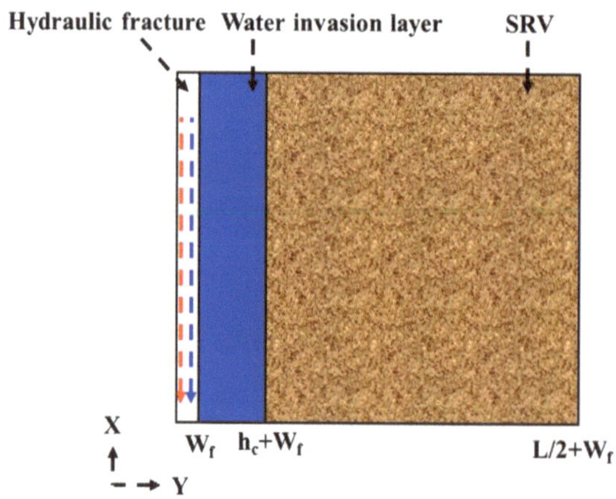

Figure 4. Gas–water two-phase flow in the major fracture.

The governing equation of gas flow in the hydraulic fracture can be expressed as follows:

$$\frac{\partial}{\partial x}(\rho_g v_F) + \rho_g q_{cg} = 0.0864 \frac{\partial(\rho_g \phi_F)}{\partial t} \tag{13}$$

where, v_{Fg} is the gas-phase fluid velocity of the hydraulic fracture, m/d; ϕ_F is hydraulic fracture porosity, dimensionless; k_F is permeability of hydraulic fractures, mD; p_F is the hydraulic fracture pressure, MPa; and q_{cg} is the gas flow rate into the hydraulic fracture per unit volume of the water invasion layer system per unit time (unit m)^3/d.

Then, the equation of gas flow in the hydraulic fracture is:

$$\begin{cases} \frac{\partial^2\psi_F}{\partial x^2} = 0.0864[\frac{\mu_{gi}(\phi c_t)_{Fi}}{k_F\hat{k}_{Frg}}\frac{\partial\psi_F}{\partial t_a} - \frac{2k_c\hat{k}_{crg}}{w_Fk_F\hat{k}_{Frg}}\frac{\partial\psi_c}{\partial y}\Big|_{y=w_F/2}] \\ \psi_F(x,t_a)|_{t_a=0} = \psi_i \\ \frac{\partial\psi_F(x,t_a)}{\partial x}\Big|_{x=x_F} = 0 \\ \psi_F(x,t_a)|_{x=0} = \psi_w \end{cases} \quad (14)$$

The water phase flow equation in the hydraulic fracture system is:

$$\begin{cases} \frac{\partial^2 p_F}{\partial x^2} = 0.0864[\frac{\phi_F c_{tF}^*\mu_w}{k_F\hat{k}_{Frw}}\frac{\partial p_F}{\partial t} - \frac{2k_c\hat{k}_{crw}}{w_Fk_F\hat{k}_{Frw}}\frac{\partial p_c}{\partial y}\Big|_{y=w_F/2}] \\ p_F(x,t)|_{t=0} = p_i \\ \frac{\partial p_F(x,t)}{\partial x}\Big|_{x=x_F} = 0 \\ p_F(x,t)|_{x=0} = p_w \end{cases} \quad (15)$$

2.5. Dimensionless Model and Lapalce Transform

For the convenience of derivation, the mathematical model adopts a dimensionless form. The definition of dimensionless is shown in Table 1:

Table 1. Dimensionless parameter definition.

Dimensionless Parameters	Define
Dimensionless x	$x_D = \frac{x}{L_r}$
Dimensionless time	$t_D = \frac{\eta_r}{L_r^2}t$
Dimensionless permeability	$k_D = \frac{k}{k_r}$
Dimensionless y	$y_D = \frac{y}{L_r}$
Dimensionless pseudo-time	$t_{aD} = \frac{\eta_r}{L_r^2}t_a$
Dimensionless hydraulic fracture conductivity	$C_{FD} = \frac{w_Fk_F}{k_cL_r} = \frac{k_{FD}w_{FD}}{k_{cD}}$

3. Solution of Gas–Water Two-Phase Flow Model

Due to the nonlinearity of the gas and water flow equations, the solution of the model is solved semi-analytically. Time is discretized, and in a single step, the average saturation is treated as a constant value. Additionally, it can have an analytic solution and be solved by Stehfest [9] to obtain the solution of the model in real space. Then, the average saturation at the next time step is calculated by using the mass balance equation of gas and water flow. Finally, the gas and water production are predicted successively, forming a semi-analytical solution method for the gas–water two-phase flow model.

3.1. Solution of Gas Flow Equation

In this paper, the solution of the model is obtained by using the Laplace transform. The dimensionless Laplace transform of each equation is taken and then can be derived. The solution of gas production is obtained as follows.

$$\overline{q}_{gD} = \frac{\hat{k}_{Frg}k_{FD}w_{FD}}{\pi}\frac{1-S_c}{s}\sqrt{c_4(s)}\tanh\left(\sqrt{c_4(s)}\cdot x_{FD}\right) \quad (16)$$

It should be noted that the gas production equation derived from analytical solution is in Laplace space, while the solution in real space needs to be obtained by a Stehfest numerical inversion algorithm. In addition, the Laplace transform is about dimensionless pseudo-time, while the Laplace transform of the water phase equation is about dimensionless time, so these two equations must be distinguished. Appendix B provides the detail derivation of the gas flow model and the constant in Equation (16) is expressed as:

$$\begin{cases} c_1(s) = \frac{k_m}{k_c}\frac{\sqrt{s/\eta_{mD}}}{\sqrt{s/(\hat{k}_{crg}\eta_{cD})}}\tanh\left[\sqrt{\frac{s}{\eta_{mD}}}(h_{cD}+w_{FD}/2-L_{FD}/2)\right] \\ c_2(s) = \cosh\left[\sqrt{\frac{s}{\hat{k}_{crg}\eta_{cD}}}(-h_{cD})\right] + c_1(s)\sinh\left[\sqrt{\frac{s}{\hat{k}_{crg}\eta_{cD}}}(-h_{cD})\right] \\ c_3(s) = \sqrt{\frac{s}{\hat{k}_{crg}\eta_{cD}}}\frac{\sinh\left[\sqrt{\frac{s}{\hat{k}_{crg}\eta_{cD}}}(-h_{cD})\right]+c_1(s)\cosh\left[\sqrt{\frac{s}{\hat{k}_{crg}\eta_{cD}}}(-h_{cD})\right]}{c_2(s)} \\ c_4(s) = \frac{s}{\hat{k}_{Frg}\eta_{FD}} - \frac{2\hat{k}_{crg}}{C_{FD}\hat{k}_{Frg}}c_3(s) \\ \overline{q}_{gD} = \frac{\hat{k}_{Frg}k_{FD}w_{FD}}{\pi}\frac{1-S_c}{s}\sqrt{c_4(s)}\tanh\left(\sqrt{c_4(s)}\cdot x_{FD}\right) \end{cases} \tag{17}$$

3.2. Solution of Water Flow Equation

Similar to the derivation of the gas flow equation in the above section, we also adopt the Laplace transform to obtain the water flow solution of the model.

The solution of water phase production is expressed as follows:

$$\overline{q}_{wD} = -\frac{\hat{k}_{Frw}k_{FD}w_{FD}}{\pi}\frac{\partial \overline{p}_{FD}}{\partial x_D}\bigg|_{x_D=0} = \frac{\hat{k}_{Frw}k_{FD}w_{FD}}{\pi}\frac{1-S_c}{s}\sqrt{d_2(s)}\tanh\left(\sqrt{d_2(s)}\cdot x_{FD}\right) \tag{18}$$

Appendix C provides the detail derivation of the gas flow model and the constant in Equation (17) is expressed as:

$$\begin{cases} d_1(s) = \sqrt{\frac{s}{\hat{k}_{crw}\eta_{cwD}}}\tanh\left[\sqrt{\frac{s}{\hat{k}_{crw}\eta_{cwD}}}(-h_{cD})\right] \\ d_2(s) = \frac{s}{\hat{k}_{Frw}\eta_{FwD}} - \frac{2\hat{k}_{crw}}{C_{FD}\hat{k}_{Frw}}d_1(s) \\ \overline{q}_{wD} = -\frac{\hat{k}_{Frw}k_{FD}w_{FD}}{\pi}\frac{1-S_c}{s}\frac{\partial \overline{p}_{FD}}{\partial x_D}\bigg|_{x_D=0} = \frac{\hat{k}_{Frw}k_{FD}w_{FD}}{\pi}\frac{1-S_c}{s}\sqrt{d_2(s)}\tanh\left(\sqrt{d_2(s)}\cdot x_{FD}\right) \end{cases} \tag{19}$$

3.3. Solution Procedure of the Model

To solve the above equations, we must rely on gas and water mass balance equations for calculating the average pressure of the SRV system and the average saturation of the fracture system, and then update the saturation-dependent and pressure-dependent parameter in Equations (16) and (18). The mass balance equation of water in the fracture and the water invasion layer are given as follows:

$$W_p = V_{pFc}\left(\frac{S_{wi}}{B_{wi}} - \frac{\hat{S}_w}{B_w}\right) \tag{20}$$

$$\hat{S}_w = S_{wi} - \frac{W_p}{V_{pFc}} \tag{21}$$

The gas mass balance equation is:

$$\begin{aligned} W_g &= V_{pm} S_{mgi}\left(\frac{1}{B_{gi}}-\frac{1}{B_g}\right)+V_{pm}\left(\frac{V_L p_i}{p_L+p_i}-\frac{V_L \hat{p}}{p_L+\hat{p}}\right) \\ &+V_{pFc}\left(\frac{1-S_{wi}}{B_{gi}}-\frac{1-\hat{S}_w}{B_g}\right)+V_{pFc}\left(\frac{V_L p_i}{p_L+p_i}-\frac{V_L \hat{p}}{p_L+\hat{p}}\right) \end{aligned} \tag{22}$$

In a shale gas reservoir, the fracture system provides a high-conductivity gas flow channel, but the compressibility of the fracture system is small. Hence, during flowback, the average pressure of the SRV system is much higher than that of the fracture system. Additionally, the average pressure of the fracture system is approximately equal to the bottom-hole flowing pressure, and the average pressure of the matrix system is close to that of the reservoir. Therefore, for the calculation of water phase flow, the unified average water saturation is used to calculate the fracture and water invasion layer, while the average pressure of the fracture system is calculated by the fracture pressure solution, and the average pressure of the water invasion layer system is calculated by the water invasion layer pressure solution. For the gas phase, the gas saturation is calculated based on the average water saturation, and the average pressure is calculated using the gas-phase mass balance equation. The model calculation process is shown in Figure 5.

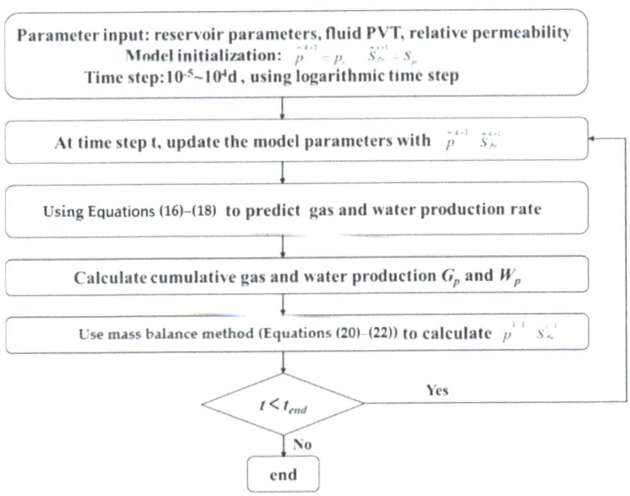

Figure 5. Semi-analytical solution flow of gas–water two-phase flow model.

3.4. Model Validation

In order to validate the accuracy of the gas–water flow model and the semi-analytical solution method, this paper uses a commercial reservoir numerical simulator to establish a set of discrete fracture systems (as shown in Figure 6). The gas–water production results of numerical simulation are compared with that of the flow model.

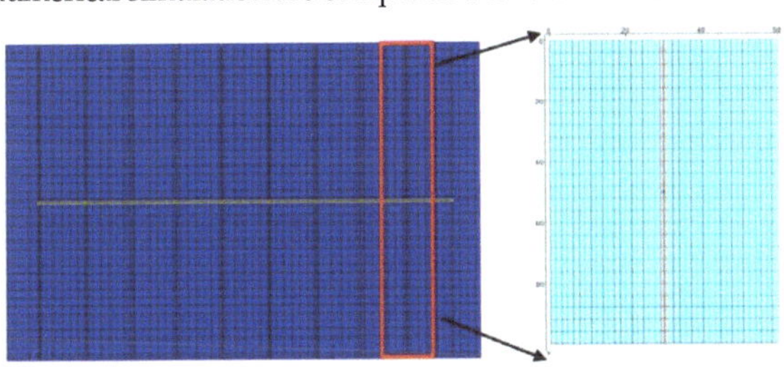

Figure 6. Schematic diagram of the numerical simulation grid.

The grid systems of the numerical simulation is 430 × 25 × 1, and the total number of grids is 10,750. In order to accurately represent the fracture width, the mesh around the fracture is refined. The mesh size near the fracture is 0.045 m and 0.01 m, and the width of the fracture is represented by the grids with the width of 0.01 m. The basic parameters used are shown in Table 2.

Table 2. Parameters used for model comparison.

Parameter	The Numerical
Fracture porosity, %	45
Crack compression coefficient, MPa^{-1}	1.0×10^{-3}
Fracture permeability, mD	1000
Crack half-length, M	100
Initial fracture water saturation, %	100
Matrix compression coefficient, MPa^{-1}	1.0×10^{-4}
Width of water invasion layer, m	2
Matrix permeability, mD	1.0×10^{-4}
Effective reservoir thickness, m	10
Initial gas saturation of the matrix, %	70
Initial crack pressure, MPa	29.5
Matrix porosity, %	6.4
Bottom hole flow pressure, MPa	2
Permeability of water invasion layer, mD	5.0×10^{-5}

The comparison results are shown in Figures 7 and 8. It can be seen that the water and gas production predicted by the two methods are in good agreement. Therefore, the established gas–water flow model is correct and can accurately characterize the gas–water flow process of shale reservoirs, and the model is relatively reliable.

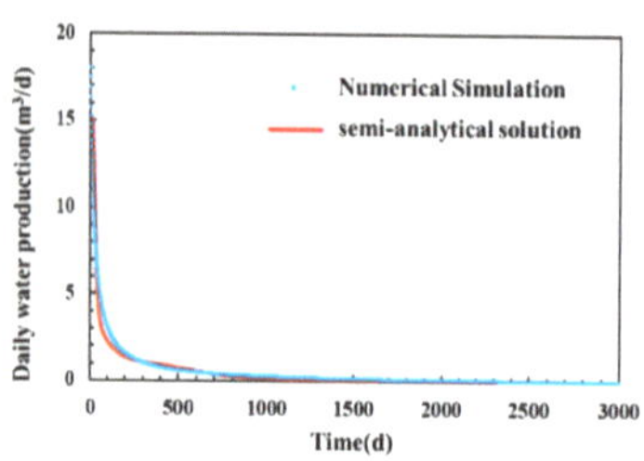

Figure 7. Daily water flow curves.

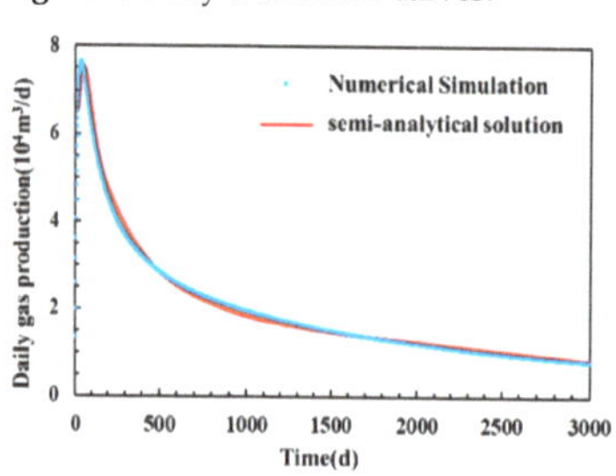

Figure 8. Daily gas volume.

4. Interpretation of Fracture Network Parameters

4.1. Flow Regimes of Gas-Water

There are usually two modes of production in shale gas reservoirs. The first mode shows gas–water two-phase flow in the early flowback period, while the second mode usually shows a single-phase fracturing fluid flowback period with several days. The gas in the matrix enters the fracture and then shows a two-phase flow.

4.1.1. Flow Regimes of Gas-Water in the First Mode

Figure 9 shows the flow characteristics of the first mode of gas–water production.

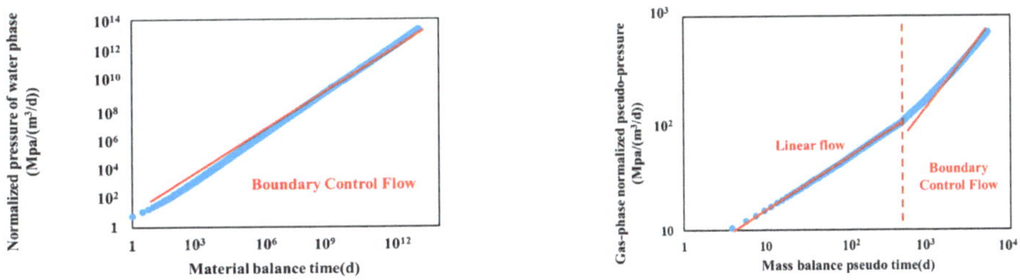

(**a**) (**b**)

Figure 9. Typical gas–water flow regimes in the first type of production mode. (**a**) Typical flow characteristics of water phase, and (**b**) typical flow characteristics of gas phase.

(1) Water boundary dominated flow. Because the fracturing fluid mainly occurs in the hydraulic fracture, water will drain very fast and then show the boundary dominated flow, until the water production rate is nearly zero. On the plot of rate normalized pressure and mass balance time, the curve shows a unit slope straight line.

(2) Gas linear flow. This flow regime occurs at almost the same time with a water boundary dominated flow, and is the first flow stage of the gas phase. At this time, the gas in the matrix flows to the fracture in a direction perpendicular to the fracture surface, which appears as straight with one-half slope on the typical curve.

(3) Gas boundary dominated flow. With the production, the gas pressure drop reaches the closed boundary and forms the boundary dominated flow, which is represented as a straight line with a unit slope on the typical curve shown.

The description and schematic diagram of each flow regimes are summarized in Table 3.

Table 3. Description of gas–water flow in the first type of production mode.

Flow Regime	Description	Schematic Diagram
Water boundary dominated flow	The fracturing fluid flows along the hydraulic fracture into the wellbore, and the water exhibits depletion in the fracture	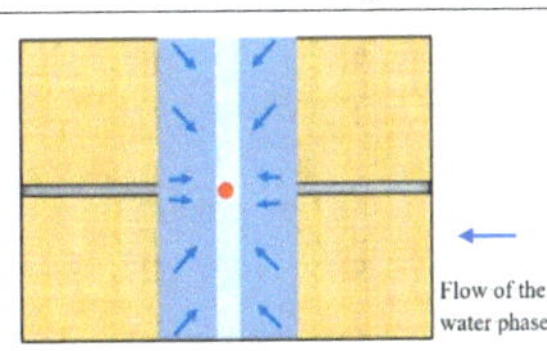
Gas linear flow	Gas in matrix flows into the fracture in a direction perpendicular to the fracture surface, and the gas dominates the fracture flow	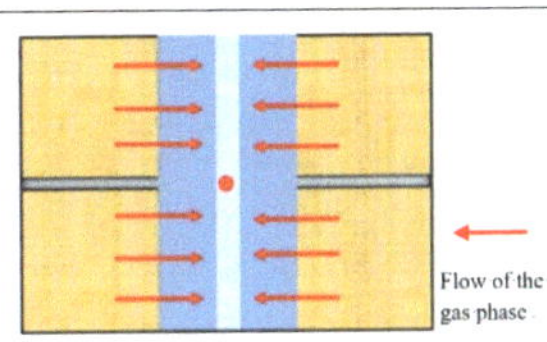
Gas boundary dominated flow	After the gas pressure drop boundary reaches the closed boundary, the closed boundary control flow is formed	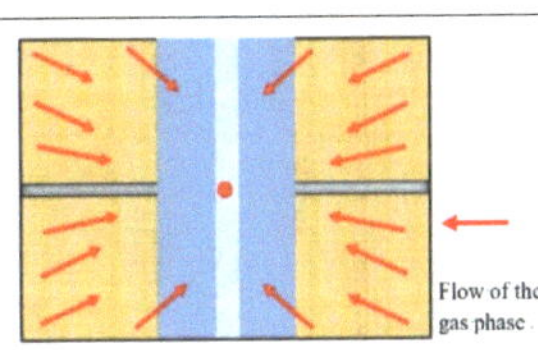

4.1.2. Flow Regimes of Gas-Water in the Second Mode

Figure 10 describes the flow characteristics in each stage of the second type of gas–water production.

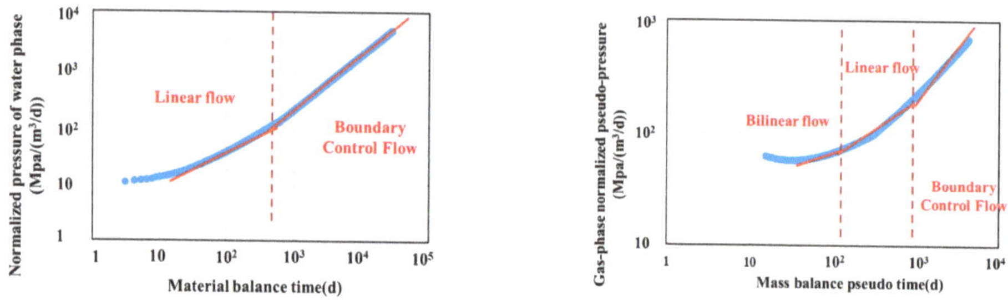

(a) (b)

Figure 10. Gas–water flow stage in the second type of production dynamic mode. (**a**) Typical flow characteristics of water phase, and (**b**) typical flow characteristics of gas phase.

(1) Water linear flow. Since fracturing fluid exists in both the hydraulic fracture and the invasion layer, linear flow of the water phase will occur when the pressure drop in the water invasion layer after the flowback starts. The typical curve shows a straight line with a unit slope.
(2) Water boundary dominated flow. Due to the fracturing fluid mainly occurs in the hydraulic fracture, water will drain very fast and then show the boundary dominated flow, until the water production rate is nearly zero. On the plot of rate normalized pressure and mass balance time, the curve shows a unit slope straight line.
(3) Gas bilinear flow. Gas phase flow into the hydraulic fracture soon after the flowback starts and forms bilinear flow with the pressure drop system formed by the water invasion layer and shale matrix, which is represented as a straight line with a quarter-slope.
(4) Gas linear flow. This flow regime occurs at almost the same time with water boundary dominated flow, and is the first flow stage of the gas phase. At this time, the gas in the matrix flows to the fracture in a direction perpendicular to the fracture surface, which appears as a straight line with one-half slope on the typical curve.
(5) Gas boundary dominated flow. With the production, the gas pressure drop reaches the closed boundary and forms the boundary dominated flow, which is represented as a straight line with a unit slope on the typical curve shown.

The description and schematic diagram of each flow regime are summarized in Table 4.

Table 4. Description of gas–water flow in the second type of production mode.

Liquid Phase	Flow Phase Description	Schematic Diagram
Water linear flow	The water fluid flows linearly to the wellbore along the invasion layer. This stage occurs at the beginning of the flowback.	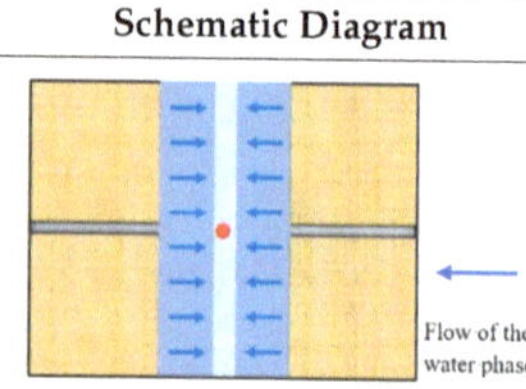

Table 4. *Cont.*

Liquid Phase	Flow Phase Description	Schematic Diagram
Water boundary dominated flow	The fracturing fluid flows along the hydraulic fracture into the wellbore, and the water exhibits depletion in the fracture	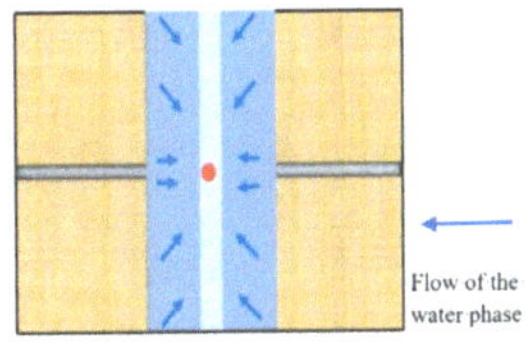
Gas linear flow	Gas in matrix flows into the fracture in a direction perpendicular to the fracture surface, and the gas dominates the fracture flow	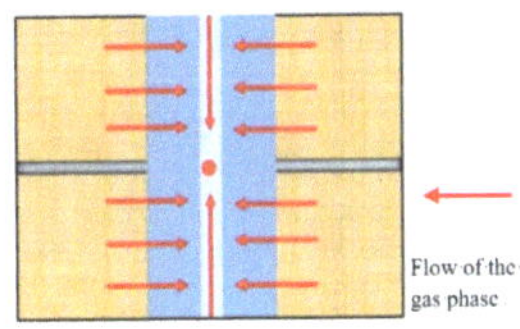
Gas boundary dominated flow	After the gas pressure drop boundary reaches the closed boundary, the closed boundary control flow is formed	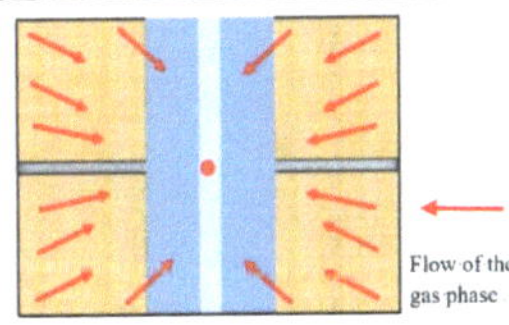

4.2. Interpretation Procedure Fracture Network Parameters

The interpretation of the fracture network is to fit the dynamic data of gas and water production using the new model. Based on the new model, this section proposes a fitting method for the problem of variable flow pressure production. Thus, Equation (23) can be used to convert the production under constant bottomhole flow pressure into that under variable flow pressure.

$$q(t) = \sum_{k=1}^{n} \left[\left(\Delta\psi_{wf,k} - \Delta\psi_{wf,k-1} \right) \cdot q_{cp}(t - t_{k-1}) \right] \tag{23}$$

where, $\left(\Delta\psi_{wf,k} - \Delta\psi_{wf,k-1} \right)$ is the pressure history and $q_{cp}(t - t_{k-1})$ is the constant pressure solution. Thus, the main steps are as follows:

(1) Initial parameter input: set the initial value of the parameters of fracture network, including the fracture half-length, the fracture permeability, the thickness of the water invasion layer, the permeability of the water invasion layer, the width and permeability of the SRV. Additionally, set $\overline{p} = p_1, \overline{S}_g = S_{g1}$.

(2) Take days as the time step, and update the parameters of the gas–water flow model with the $\overline{p}, \overline{S}_g, \eta, \alpha, \beta, \chi$.

(3) Constant pressure solution: gas and water production are calculated by using the gas–water two-phase flow model.

(4) Variable pressure solution: Equation (23) is used to obtain the gas production and water production under variable bottomhole flow pressure.

(5) Calculation of average pressure and saturation: the average pressure and saturation calculated by the material balance method are then substituted into step (2) for recurrent calculation.

(6) If the fitting error between simulated and actual data meets the criterion of convergence, the calculation is finished. Otherwise, update the fitting parameters and repeat the above steps.

The detailed flow chart is shown in Figure 11.

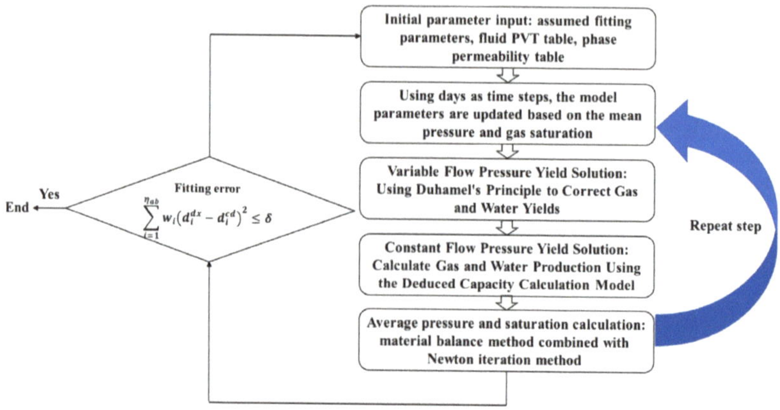

Figure 11. The flow chart of interpretation of the fracture network parameters.

4.3. Field Exmaple

4.3.1. Well A

The gas–water production and bottomhole pressure of the well are shown in Figure 12.

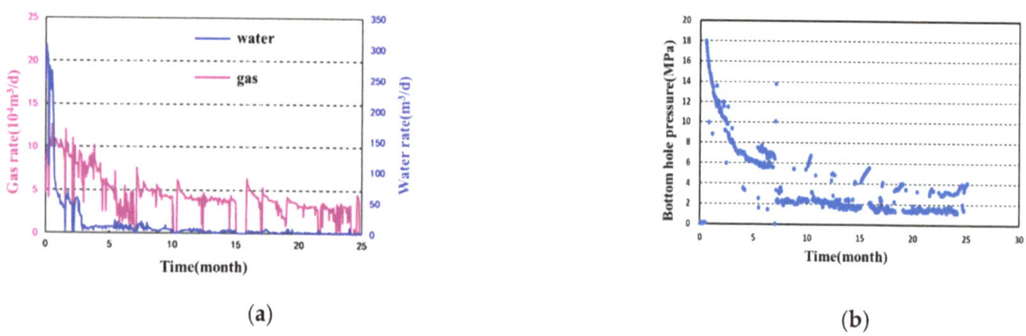

Figure 12. Gas–water production and pressure data of well A. (**a**) Gas–water production rate, and (**b**) Bottomhole flow pressure.

The reservoir parameters, fluid parameters and fracturing parameters are shown in Table 5. Analysis of gas–water two-phase flow regime of well A. are as shown in Figure 13.

Table 5. Values of basic parameters of well A.

Parameter	Value
Initial pressure, MPa	29.65
Initial gas saturation, decimal	0.6
Reservoir temperature, K	358.1
Reservoir thickness, m	15
Langmuir volume	2.86
Langmuir pressure, MPa	9.18
Horizontal well length, m	1441
Number of fractures	25
Matrix porosity	0.064
Matrix Compressibility, MPa^{-1}	8×10^{-5}
Fracture Compressibility, MPa^{-1}	8×10^{-5}
Fracture width, m	0.5×10^{-2}

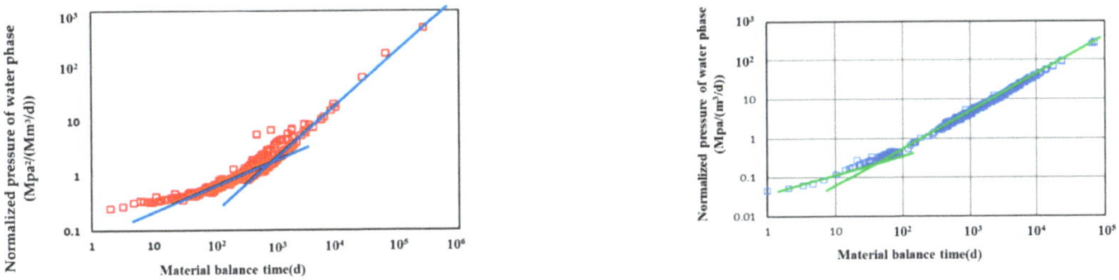

(a) (b)

Figure 13. Analysis of gas–water two-phase flow regime of well A. (**a**) Gas phase flow regime, and (**b**) water phase flow regime.

In early production, gas exhibits a long linear flow and the gas enters the fracture after 14 months. For the water, it exhibits a fracture linear flow and then the boundary dominated flow, when most fracturing fluid has been discharged. This means that the fracture conductivity is high and the fracture half-length is large.

Based on the results from Figure 14, the total length of the discrete fracture is 4105 m, and the stimulated reservoir volume is 1286×10^4 m^3. The detailed parameters of the fracture network are shown in Table 6.

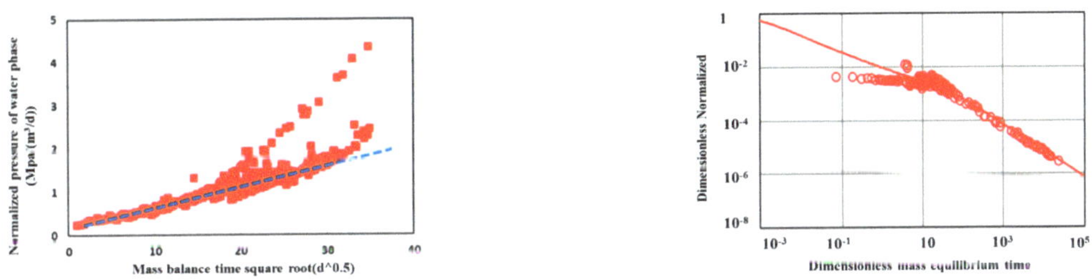

(a) (b)

Figure 14. Production decline analysis of gas phase of well B with gas–water two phases. (**a**) Gas phase linear flow analysis, and (**b**) gas production data matching.

Table 6. Interpretation of fracture network parameters of well A.

Interpreted Parameters	Value
Fracture half-length (m)	83.1
Fracture permeability (mD)	865
Thickness of water invasion layer (m)	0.12
Permeability of water invasion (mD)	2.82×10^{-3}
SRV width (m)	46.5
SRV Permeability (mD)	3.87×10^{-3}

4.3.2. Well B

The gas–water production and bottomhole pressure of the well are shown in Figure 15.

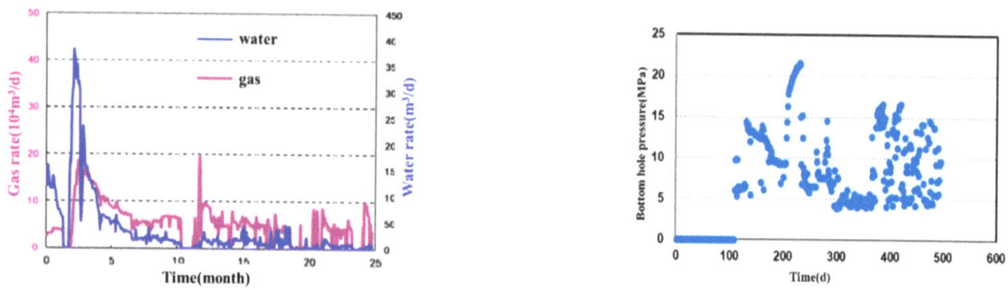

(a) (b)

Figure 15. Gas–water production and pressure data of well A. (**a**) Gas–water production rate, and (**b**) bottomhole flow pressure.

The reservoir parameters, fluid parameters and fracturing parameters are shown in Table 7. Analysis of gas–water two-phase flow stage in well 8 are as shown in Figure 16.

Table 7. Values of basic parameters of well B.

Parameter	Value
Initial pressure, MPa	31.15
Initial gas saturation, decimal	0.6
Reservoir temperature, K	383.1
Reservoir thickness, m	10
Langmuir volume	1.96
Langmuir pressure, MPa	2200
Horizontal well length, m	1266
Number of fractures	19
Matrix porosity	0.064
Matrix Compressibility, MPa^{-1}	8×10^{-5}
Fracture Compressibility, MPa^{-1}	8×10^{-5}
Fracture width, m	0.5×10^{-2}

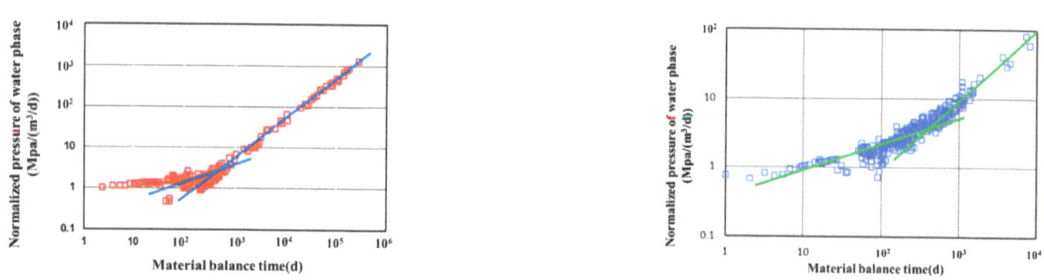

(**a**) Gas flow regime (**b**) Water flow regime

Figure 16. Analysis of gas–water two-phase flow stage in well 8. (**a**) Gas flow regime, and (**b**) water flow regime.

In the early production, the gas linear flow was long, and after 2 months the gas phase entered the fracture. For the water, it experiences fracture linear flow, and the water enters the boundary dominated flow and most fracturing fluid has been discharged. Thus, the fracture conductivity is high and the fracture half-length is small. The permeability of the SRV is large.

From Figure 17, the total length of the discrete fracture is 5152 m, and the stimulated reservoir volume is 233 × 10^4 m^3. The interpreted fracture network parameters are shown in Table 8.

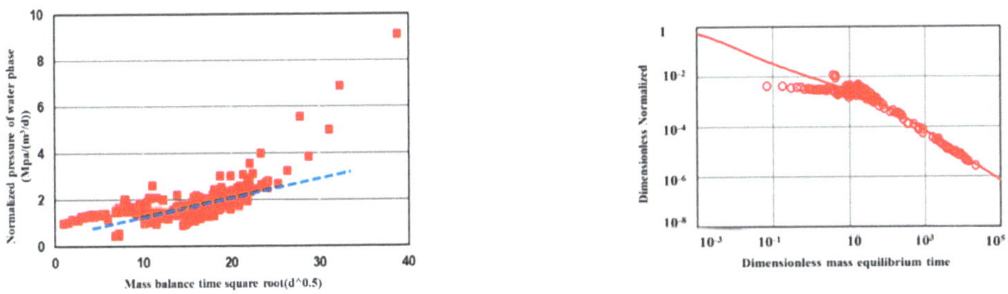

(a) (b)

Figure 17. Production decline analysis of gas phase under gas–water two phases in well B. (**a**) Gas linear flow analysis, and (**b**) gas phase production data matching.

Table 8. Interpretation of fracture network parameters of well B.

Interpreted Parameters	Value
Fracture half-length (m)	82.1
Fracture permeability (mD)	685
Thickness of water invasion layer (m)	0.91
Permeability of water invasion (mD)	1.95 × 10^{-3}
SRV width (m)	45.3
SRV Permeability (mD)	3.12 × 10^{-3}

5. Conclusions

1. In this paper, a practical model considering gas adsorption of the shale matrix, water invasion layer and gas–water two-phase flow of hydraulic fracture is established, and the semi-analytical solution method is developed.
2. The new model is used to successfully analyze the production performance of the two modes of gas and water production. The two cases show different gas and water flow regimes:
    (1) The first production mode mainly shows three flow regimes, including water boundary dominated flow, gas linear flow and gas boundary dominated flow.
    (2) The second production mode mainly shows five flow stages, including water linear flow, water boundary dominated flow, gas bilinear flow, gas linear flow and gas boundary dominated flow.
3. The developed fracture parameters interpretation method can reasonably estimate the key parameters of the hydraulic fracture, water invasion layer and SRV of fracture well from the field example.

Author Contributions: Conceptualization, P.J. and L.N.; methodology, P.J.; software, Y.L.; validation, L.N., Y.L. and H.F.; formal analysis, P.J.; investigation, P.J. and L.N.; data curation, L.N. and Y.L.; writing—original draft preparation, L.N. and H.F.; writing—review and editing, P.J.; supervision, P.J.; project administration, P.J.; funding acquisition, P.J. All authors have read and agreed to the published version of the manuscript.

Funding: This research was partially funded by the National Natural Science Foundation of China (No. 52004307). We would further like to thank the financial support of the Science and technology project of CNPC-major project (No. ZLZX2020-02-04).

Data Availability Statement: Not applicable.

Conflicts of Interest: The authors declare no conflict of interest. The funders had no role in the design of the study; in the collection, analyses, or interpretation of data; in the writing of the manuscript; or in the decision to publish the results.

Appendix A

The flow equation of the SRV can be written as:

$$\frac{\partial}{\partial y}(-\rho_{sc}v_m) = 0.0864[\frac{\partial(\rho\phi_m)}{\partial t} + \frac{\partial V}{\partial t}] \tag{A1}$$

Introducing the gas isothermal adsorption equation:

$$V = \rho_{sc}\frac{V_L p_m}{p_L + p_m} \tag{A2}$$

Since the gas flow in SRV is regarded as an isothermal flow, dimensionless gas compression factor Z is introduced, and the gas phase isothermal compression coefficient is defined as:

$$c_g(p_m) = \frac{1}{p_m} - \frac{1}{z}\frac{dz}{dp_m} \tag{A3}$$

Gas density:

$$\rho_{sc} = \frac{pM}{zRT} \tag{A4}$$

Define the matrix compressibility as:

$$c_{pm} = \frac{1}{\phi_m}\frac{d\phi_m}{dp_m} \tag{A5}$$

Introduce the pseudo-pressure:

$$\psi_m(p_m) = 2\int_{p_i}^{p_m}\frac{p}{\mu_g z}dp \tag{A6}$$

Since gas flows isothermal in the SRV, the equation can be simplified to:

$$\frac{\partial^2\psi_m}{\partial x^2} = 0.0864\frac{\phi_m\mu(p_m)}{k_m}[c_{pm} + c_g(p_m) + \frac{z(p_m)p_{sc}T}{\phi_m z_{sc}T_{sc}}\frac{V_L p_L}{p_m(p_L + p_m)^2}]\frac{\partial\psi_m}{\partial t} \tag{A7}$$

where,

$$c_{tm}(p_m) = c_{pm} + c_g(p_m) + \frac{z(p_m)p_{sc}T}{\phi_m z_{sc}T_{sc}}\frac{V_L p_L}{p_m(p_L + p_m)^2} \tag{A8}$$

The last term in the above Equation (A8) represents the effect of desorption, so it is defined as the desorption compressibility C_d, the unit MPa^{-1}.

$$c_d(p_m) = \frac{z(p_m)p_{sc}T}{\phi_m z_{sc}T_{sc}}\frac{V_L p_L}{p_m(p_L + p_m)^2} \tag{A9}$$

Usually, an average formation pressure is adopted to replace the matrix pressure in the above equation, and take the average pressure as a constant value:

$$c_d = \frac{p_{sc}z(\overline{p}_m)TV_L p_L}{\phi_m z_{sc}T_{sc}\overline{p}_m(p_L + \overline{p}_m)^2} \tag{A10}$$

Therefore, the flow equation in the SRV is expressed as

$$\frac{\partial^2 \psi_m}{\partial y^2} = 0.0864\frac{\phi_m \mu_g(p_m)c_{tm}(p_m)}{k_m}\frac{\partial \psi_m}{\partial t} \tag{A11}$$

Furthermore, for the nonlinear term on the right side of the equation, pseudo-time is used, which is defined as:

$$t_a = \int_0^t \frac{\mu_{gi}c_{tmi}}{\mu_g(\overline{p}_m)c_{tm}(\overline{p}_m)}dt \tag{A12}$$

Thus, the single-phase gas flow equation in the SRV is changed by use of pseudo-pressure and pseudo-time, and the flow equation can be obtained as follows:

$$\begin{cases} \frac{\partial^2 \psi_m}{\partial y^2} = \frac{\mu_{gi}(\phi c_t)_{mi}}{0.0864k_m}\frac{\partial \psi_m}{\partial t_a} \\ \psi_m(y,t)|_{t_a=0} = \psi_i \\ \left.\frac{\partial \psi_m(y,t_a)}{\partial y}\right|_{y=L_F/2} = 0 \end{cases} \tag{A13}$$

The inner boundary of this region is coupled with the water invasion layer, and the pressure and flow rate on the inner boundary and the water invasion layer should be equal, respectively:

$$\psi_m|_{y=h_c+w_F/2} = \psi_c|_{y=h_c+w_F/2} \tag{A14}$$

Appendix B

Derivation of the flow equation of system in SRV. The solution of the model is obtained by using the Laplace transform.

For the SRV, the above equation is:

$$\frac{\partial^2 \overline{\psi}_{mD}}{\partial y_D^2} = \frac{s}{\eta_{mD}}\overline{\psi}_{mD} \tag{A15}$$

The general solution of Equation (A15) is:

$$\overline{\psi}_{mD} = A_2 \cosh\left[\sqrt{\frac{s}{\eta_{mD}}}(y_D - L_{FD}/2)\right] + B_2 \sinh\left[\sqrt{\frac{s}{\eta_{mD}}}(y_D - L_{FD}/2)\right] \tag{A16}$$

Substituting the initial and outer boundary conditions, B will be zero. Using the inner boundary conditions, we can obtain:

$$A_2 = \frac{\psi_{cD}(h_{cD} + w_{FD}/2)}{\cosh\left[\sqrt{\frac{s}{\eta_{mD}}}(h_{cD} + w_{FD}/2 - L_{FD}/2)\right]} \tag{A17}$$

Thus, the pressure solution in the SRV is:

$$\overline{\psi}_{mD}(y_D) = \psi_{cD}(h_{cD} + w_{FD}/2)\frac{\cosh\left[\sqrt{\frac{s}{\eta_{mD}}}(y_D - L_{FD}/2)\right]}{\cosh\left[\sqrt{\frac{s}{\eta_{mD}}}(h_{cD} + w_{FD}/2 - L_{FD}/2)\right]} \tag{A18}$$

Then, the flow between the SRV and the water invasion layer can be written as:

$$\begin{aligned}&\frac{k_m}{\mu}\frac{\partial\psi_{mD}}{\partial y_D}\bigg|_{y_D=h_{cD}+w_{FD}/2}\\&=\frac{k_m}{\mu}\psi_{cD}(h_{cD}+w_{FD}/2)\sqrt{\frac{s}{\eta_{mD}}}\tanh\left[\sqrt{\frac{s}{\eta_{mD}}}(h_{cD}+w_{FD}/2-L_{FD}/2)\right]\end{aligned} \tag{A19}$$

For the water invasion layer, the general solution of the flow model can be obtained as follows:

$$\overline{\psi}_{cD}(y_D)=A_1\cosh\left[\sqrt{\frac{s}{\hat{k}_{crg}\eta_{cD}}}(y_D-h_{cD}-w_{FD}/2)\right]+B_1\sinh\left[\sqrt{\frac{s}{\hat{k}_{crg}\eta_{cD}}}(y_D-h_{cD}-w_{FD}/2)\right] \tag{A20}$$

According to the boundary condition of the SRV, we have

$$\begin{aligned}B_1&=\psi_{cD}(h_{cD}+w_{FD}/2)\frac{k_m}{k_c}\frac{\sqrt{s/\eta_{mD}}}{\sqrt{s/(\hat{k}_{crg}\eta_{cD})}}\tanh\left[\sqrt{\frac{s}{\eta_{mD}}}(h_{cD}+w_{FD}/2-L_{FD}/2)\right]\\&=\psi_{cD}(h_{cD}+w_{FD}/2)c_1(s)\end{aligned} \tag{A21}$$

Type,

$$c_1(s)=\frac{k_m}{k_c}\frac{\sqrt{s/\eta_{mD}}}{\sqrt{s/\left(\hat{k}_{crg}\eta_{cD}\right)}}\tanh\left[\sqrt{\frac{s}{\eta_{mD}}}(h_{cD}+w_{FD}/2-L_{FD}/2)\right] \tag{A22}$$

Then, the general solution of the flow model of the water invasion layer can be written as follows:

$$\overline{\psi}_{cD}(y_D)=\psi_{cD}(h_{cD}+w_{FD}/2)\left\{\begin{array}{c}\cosh\left[\sqrt{\frac{s}{\hat{k}_{crg}\eta_{cD}}}(y_D-h_{cD}-w_{FD}/2)\right]\\+c_1(s)\sinh\left[\sqrt{\frac{s}{\hat{k}_{crg}\eta_{cD}}}(y_D-h_{cD}-w_{FD}/2)\right]\end{array}\right\} \tag{A23}$$

$$\begin{aligned}\psi_{cD}(h_{cD}+w_{FD}/2)&=\frac{\overline{\psi}_{FD}}{\cosh\left[\sqrt{\frac{s}{\hat{k}_{crg}\eta_{cD}}}(-h_{cD})\right]+c_1(s)\sinh\left[\sqrt{\frac{s}{\hat{k}_{crg}\eta_{cD}}}(-h_{cD})\right]}\\&=\frac{\overline{\psi}_{FD}}{c_2(s)}\end{aligned} \tag{A24}$$

where,

$$c_2(s)=\cosh\left[\sqrt{\frac{s}{\hat{k}_{crg}\eta_{cD}}}(-h_{cD})\right]+c_1(s)\sinh\left[\sqrt{\frac{s}{\hat{k}_{crg}\eta_{cD}}}(-h_{cD})\right] \tag{A25}$$

Thus, the pressure solution of the water invasion layer is:

$$\overline{\psi}_{cD}(y_D)=\overline{\psi}_{FD}\frac{\cosh\left[\sqrt{\frac{s}{\hat{k}_{crg}\eta_{cD}}}(y_D-h_{cD}-w_{FD}/2)\right]+c_1(s)\sinh\left[\sqrt{\frac{s}{\hat{k}_{crg}\eta_{cD}}}(y_D-h_{cD}-w_{FD}/2)\right]}{c_2(s)} \tag{A26}$$

Then, the flow rate of water invasion layer into the hydraulic fracture is:

$$\frac{\partial\overline{\psi}_{cD}}{\partial y_D}\bigg|_{y_D=w_{FD}/2}=\overline{\psi}_{FD}\sqrt{\frac{s}{\hat{k}_{crg}\eta_{cD}}}\frac{\sinh\left[\sqrt{\frac{s}{\hat{k}_{crg}\eta_{cD}}}(-h_{cD})\right]+c_1(s)\cosh\left[\sqrt{\frac{s}{\hat{k}_{crg}\eta_{cD}}}(-h_{cD})\right]}{c_2(s)}=\overline{\psi}_{FD}c_3(s) \tag{A27}$$

where,

$$c_3(s) = \sqrt{\frac{s}{\hat{k}_{crg}\eta_{cD}}} \frac{\sinh\left[\sqrt{\frac{s}{\hat{k}_{crg}\eta_{cD}}}(-h_{cD})\right] + c_1(s)\cosh\left[\sqrt{\frac{s}{\hat{k}_{crg}\eta_{cD}}}(-h_{cD})\right]}{c_2(s)} \tag{A28}$$

For the hydraulic fracture, it can be obtained by substituting the flow rate formula of water invasion layer to fracture

$$\frac{\partial^2\overline{\psi}_{FD}}{\partial x_D^2} + \frac{2\hat{k}_{crg}}{C_{FD}\hat{k}_{Frg}}\overline{\psi}_{FD}c_3(s) - \frac{s}{\hat{k}_{Frg}\eta_{FD}}\overline{\psi}_{FD} = 0 \tag{A29}$$

The above equation can be written as:

$$\frac{\partial^2\overline{\psi}_{FD}}{\partial x_D^2} - c_4(s)\overline{\psi}_{FD} = 0 \tag{A30}$$

$$c_4(s) = \frac{s}{\hat{k}_{Frg}\eta_{FD}} - \frac{2\hat{k}_{crg}}{C_{FD}\hat{k}_{Frg}}c_3(s) \tag{A31}$$

Similarly, the general solution of the hydraulic fracture flow model can be obtained as follows:

$$\overline{\psi}_{FD}(x_D) = A_F\cosh\left[\sqrt{c_4(s)}(x_D - x_{FD})\right] + B_F\sinh\left[\sqrt{c_4(s)}(x_D - x_{FD})\right] \tag{A32}$$

By using closed outer boundary condition, we find that B_F equals to zero. Additionally, A_F can be obtained by the use of constant pressure inner boundary condition

$$A_F = \frac{1 - S_c}{s}\frac{1}{\cosh\left[\sqrt{c_4(s)}(-x_{FD})\right]} \tag{A33}$$

Thus, the hydraulic fracture flow model is solved as follows:

$$\overline{\psi}_{FD}(x_D) = \frac{1 - S_c}{s}\frac{\cosh\left[\sqrt{c_4(s)}(x_D - x_{FD})\right]}{\cosh\left[\sqrt{c_4(s)}(-x_{FD})\right]} \tag{A34}$$

The gas production of the horizontal well through hydraulic fracture can be expressed as:

$$\overline{q}_{gD} = -\frac{\hat{k}_{Frg}k_{FD}w_{FD}}{\pi}\left.\frac{\partial\overline{\psi}_{FD}}{\partial x_D}\right|_{x=0} \tag{A35}$$

Substituting the solution of the hydraulic fracture model, the solution of gas production is:

$$\overline{q}_{gD} = \frac{\hat{k}_{Frg}k_{FD}w_{FD}}{\pi}\frac{1 - S_c}{s}\sqrt{c_4(s)}\tanh\left(\sqrt{c_4(s)}\cdot x_{FD}\right) \tag{A36}$$

It should be noted that the gas production derived from analytical solution is in Laplace space, while the solution in real space needs to be obtained by a Stehfest numerical inversion algorithm.

Appendix C

Steps for solving the water phase flow equation. For the water invasion layer, the above equation can be changed into:

$$\frac{\partial^2 \overline{p}_{cD}}{\partial y_{cD}^2} = \frac{s}{\hat{k}_{crw}\eta_{cwD}}\frac{\partial \overline{p}_{cD}}{\partial t_D} \tag{A37}$$

The general solution of Equation (A38) is as follows.

$$\overline{p}_{cD} = A_1 \cosh\left[\sqrt{\frac{s}{\hat{k}_{crw}\eta_{cwD}}}(y_D - h_{cD} - w_{FD}/2)\right] + B_1 \sinh\left[\sqrt{\frac{s}{\hat{k}_{crw}\eta_{cwD}}}(y_D - h_{cD} - w_{FD}/2)\right] \tag{A38}$$

Substituting the initial and outer boundary conditions, we find that B equals to zero. By use of the inner boundary condition, we have

$$A_1 = \frac{\overline{p}_{FD}}{\cosh\left[\sqrt{\frac{s}{\hat{k}_{crw}\eta_{cwD}}}(-h_{cD})\right]} \tag{A39}$$

Thus, the pressure distribution of water invasion layer is

$$\overline{p}_{cD}(y_D) = \overline{p}_{FD}\frac{\cosh\left[\sqrt{\frac{s}{\hat{k}_{crw}\eta_{cwD}}}(y_D - h_{cD} - w_{FD}/2)\right]}{\cosh\left[\sqrt{\frac{s}{\hat{k}_{crw}\eta_{cwD}}}(-h_{cD})\right]} \tag{A40}$$

Further, the water flow from the water invasion layer to hydraulic fracture can be obtained as follows:

$$\left.\frac{\partial \overline{p}_{cD}}{\partial y_D}\right|_{y_D = w_{FD}/2} = \overline{p}_{FD}\sqrt{\frac{s}{\hat{k}_{crw}\eta_{cwD}}}\tanh\left[\sqrt{\frac{s}{\hat{k}_{crw}\eta_{cwD}}}(-h_{cD})\right] = \overline{p}_{FD} d_1(s) \tag{A41}$$

where,

$$d_1(s) = \sqrt{\frac{s}{\hat{k}_{crw}\eta_{cwD}}}\tanh\left[\sqrt{\frac{s}{\hat{k}_{crw}\eta_{cwD}}}(-h_{cD})\right] \tag{A42}$$

For the hydraulic fracture, the flow model becomes:

$$\frac{\partial^2 \overline{p}_{FD}}{\partial x_D^2} = \frac{s}{\hat{k}_{Frw}\eta_{FwD}}\overline{p}_{FD} - \frac{2\hat{k}_{crw}}{C_{FD}\hat{k}_{Frw}}\overline{p}_{FD} d_1(s) \tag{A43}$$

The above equation can be written as:

$$\frac{\partial^2 \overline{p}_{FD}}{\partial x_D^2} - d_2(s)\overline{p}_{FD} = 0 \tag{A44}$$

$$d_2(s) = \frac{s}{\hat{k}_{Frw}\eta_{FwD}} - \frac{2\hat{k}_{crw}}{C_{FD}\hat{k}_{Frw}} d_1(s) \tag{A45}$$

Similarly, the general solution of the flow model of the hydraulic fracture can be obtained as follows:

$$\overline{p}_{FD}(x_D) = A_F \cosh\left[\sqrt{d_2(s)}(x_D - x_{FD})\right] + B_F \sinh\left[\sqrt{d_2(s)}(x_D - x_{FD})\right] \tag{A46}$$

By use of a closed outer boundary, we find that B_F equals to zero. Additionally, by use of the constant pressure inner boundary condition, we have

$$A_F = \frac{1 - S_c}{s} \frac{1}{\cosh\left[\sqrt{d_2(s)}(-x_{FD})\right]} \tag{A47}$$

Thus, the solution of the flow model of the hydraulic fracture is as follows:

$$\overline{p}_{FD}(x_D) = \frac{1 - S_c}{s} \frac{\cosh\left[\sqrt{d_2(s)}(x_D - x_{FD})\right]}{\cosh\left[\sqrt{d_2(s)}(-x_{FD})\right]} \tag{A48}$$

The water production obtained is

$$\overline{q}_{wD} = -\frac{\hat{k}_{Frw} k_{FD} w_{FD}}{\pi} \frac{\partial \overline{p}_{FD}}{\partial x_D}\bigg|_{x_D=0} = \frac{\hat{k}_{Frw} k_{FD} w_{FD}}{\pi} \frac{1 - S_c}{s} \sqrt{d_2(s)} \tanh\left(\sqrt{d_2(s)} \cdot x_{FD}\right) \tag{A49}$$

References

1. Wattenbarger, R.; El-Banbi, A.; Villegas, M.; Maggard, J.B. Production analysis of linear flow into fractured tight gas wells. In Proceedings of the SPE Rocky Mountain Regional/Low-Permeability Reservoirs Symposium, Denver, CO, USA, 5–8 April 1998.
2. Bello, R.O.; Wattenbarger, R.A. Modelling and Analysis of Shale Gas Production with a Skin Effect. *J. Can. Pet. Technol.* **2010**, *49*, 37–48. [CrossRef]
3. Alahmadi, H.A.H.; Wattenbarger, R.A. Triple-porosity models: One further step towards capturing fractured reservoirs heterogeneity. In Proceedings of the SPE/DGS Saudi Arabia Section Technical Symposium and Exhibition, Al-Khobar, Saudi Arabia, 15–18 May 2011.
4. Brown, M.L.; Ozkan, E.; Raghavan, R.S.; Kazemi, H. Practical Solutions for Pressure-Transient Responses of Fractured Horizontal Wells in Unconventional Shale Reservoirs. *SPE Reserv. Eval. Eng.* **2009**, *14*, 663–676. [CrossRef]
5. Stalgorova, K.; Mattar, L. Analytical Model for Unconventional Multifractured Composite Systems. *SPE Reserv. Eval. Eng.* **2013**, *16*, 246–256. [CrossRef]
6. Shuang, A. Productivity Evaluation and Production Rule of Fractured Horizontal Wells in Shale Gas. Master's Thesis, China University of Petroleum, Beijing, China, 2015.
7. AlQuaimi, B.I.; Rossen, W.R. Capillary Desaturation Curve for Residual Nonwetting Phase in Natural Fractures. *SPE J.* **2018**, *23*, 788–802. [CrossRef]
8. Chen, Y.; Fu, L.; Hao, M. Derivation and application of gas adsorption equation and desorption equation. *China Offshore Oil Gas* **2018**, *30*, 85–89.
9. Stehfest, H. Numerical Estimation of Laplace Transforms. *ACM Commun.* **1970**, *13*, 47–49. [CrossRef]

energies

MDPI

Article

Molecular Simulation Study of Gas–Water Adsorption Behavior and Mobility Evaluation in Ultra-Deep, High-Pressure Fractured Tight Sandstone Reservoirs

Yongfu Liu [1,2,3], Xuehao Pei [1,*], Fenglai Yang [1,2,3], Junjie Zhong [4,5,*], Li Dai [1,6], Cuili Wang [1,6,7], Tingya Zhou [1], Yijia Li [1] and Sa Xiao [1]

1 Tarim Oilfield Company, PetroChina, Korla 841000, China; shuai22loogn@yeah.net (Y.L.); yangfl-tlm@petrochina.com.cn (F.Y.); daili1-tlm@petrochina.com.cn (L.D.); wangcl-tlm@petrochina.com.cn (C.W.); zhouty-tlm@petrochina.com.cn (T.Z.); liyij-txn@petrochina.com.cn (Y.L.); xiaos-txn@petrochina.com.cn (S.X.)
2 R&D Center for Ultra-Deep Complex Reservoir Exploration and Development, China National Petroleum Corporation, Korla 841000, China
3 Engineering Research Center for Ultra-Deep Complex Reservoir Exploration and Development, Xinjiang Uygur Autonomous Region, Korla 841000, China
4 National Key Laboratory of Deep Oil and Gas, China University of Petroleum (East China), Qingdao 266580, China
5 School of Petroleum Engineering, China University of Petroleum (East China), Qingdao 266555, China
6 Xinjiang Key Laboratory of Ultra-Deep Oil and Gas, Korla 841000, China
7 Key Laboratory of Gas Reservoir Formation and Development, China National Petroleum Corporation, Korla 841000, China
* Correspondence: peixh-tlm@petrochina.com.cn (X.P.); zhongjunjie@upc.edu.cn (J.Z.)

Abstract: Under high-temperature and high-pressure conditions, understanding the competitive adsorption and mobilization mechanisms of gas and water in fractured tight sandstone gas reservoirs is crucial for optimizing the recovery factor. This study employs molecular dynamics simulation to investigate the adsorption behavior and mobilization characteristics of H_2O and CH_4 in 10 nm quartz nanopores under the conditions of the Keshen fractured tight sandstone gas reservoir. The results indicate that H_2O exhibits strong adsorption on the quartz surface, forming two high-density adsorption layers with a thickness of approximately 0.6 nm, whereas CH_4 forms three adsorption layers with a thickness of about 1.1 nm. Under gas–water coexistence conditions, the competitive adsorption effect of the water phase significantly influences the distribution of CH_4. Due to the hydrophilicity of the quartz wall, H_2O molecules preferentially adsorb onto the wall surface, forming a stable water film that significantly inhibits CH_4 adsorption. When the water saturation reaches 35%, water molecules form liquid bridges within the pores, segmenting the gas phase into different regions. As water saturation further increases, more stable liquid bridge structures develop, and microscopic water lock effects emerge, further restricting gas flow. During depletion development, H_2O remains difficult to mobilize due to strong wall adsorption, with a recovery factor of only 7%. In contrast, CH_4 exhibits high mobility, with a recovery factor of up to 75%. However, as water saturation increases from 30% to 70%, the recovery factor of CH_4 decreases significantly from 75% to 29%, indicating that the water phase not only occupies pore space, but also exerts a blocking effect that significantly inhibits CH_4 percolation and production. This study provides important theoretical support for the development strategies of ultra-deep fractured tight sandstone gas reservoirs and offers key insights for improving the ultimate recovery factor under gas–water coexistence conditions.

Keywords: ultra-deep fractured tight sandstone gas reservoirs; gas–water two-phase; nanopores; molecular simulation

1. Introduction

With the exploitation of shallow-to-middle oil and gas resources, the exploration and development of ultra-deep oil and gas have gradually become the focus and hotspot [1–3]. In China, significant breakthroughs have been made in deep reservoir exploration, particularly in the Keshen gas field, which has emerged as a key area for increasing natural gas reserves and production [4–6]. The Keshen gas field, located in the Keshen structural belt of the Kuqa foreland basin at the northern Tarim Basin, is one of the primary regions contributing to natural gas production growth. However, it is characterized as an ultra-deep, ultra-high-pressure fractured tight sandstone gas reservoir [7,8], with extreme burial depths (5500–8100 m), high formation temperatures (106–175 °C), and pressures ranging from 88.9 to 150 MPa. In addition, its strong heterogeneity and complex gas–water interactions pose significant challenges to conventional development strategies. In particular, under high-temperature, high-pressure, and confined-space conditions, gas adsorption behavior and water phase blockage can significantly impact permeability and the ultimate recovery factor [9,10]. Therefore, a comprehensive understanding of gas–water transport and competitive adsorption at the microscale is crucial for the efficient development of the Keshen gas field.

In recent years, advancements in microscale experimental techniques and computational simulation methods have highlighted the critical influence of confined spaces on gas adsorption behavior. Researchers have employed molecular dynamics simulations and other approaches to explore this phenomenon in greater detail. Zhang et al. used molecular dynamics simulation to investigate CH_4 adsorption behavior in confined spaces, focusing on nanopore adsorption layer formation. They analyzed the effects of pressure, pressure gradient, pore width, and temperature on adsorption. The results showed that as pressure increases from 1 MPa to 80 MPa, the CH_4 adsorption layer transitions from a single layer to three layers, forming a multilayer adsorption structure. Although higher pressure enhances CH_4 molecular interactions, the force exerted by the pore wall remains unchanged [11]. Ren et al. studied methane flow behavior in nanopores using molecular dynamics simulation. They developed a slit pore model to examine the effects of pore size, pressure, mineral composition, and pore water saturation on methane diffusion. Their findings indicate that methane diffusion accelerates with increasing temperature and pore size but slows as pressure rises [12]. Zhang investigated the transport behavior of natural gas and oil in a single nanopore under reservoir conditions using molecular dynamics. The study examined the behavior of $C_{10}H_{22}$, CO_2, and CH_4 in a 4 nm nanopore. The results showed that CH_4 and CO_2 form distinct adsorbed and free molecular groups, leading to different extraction behaviors. However, both gases follow similar diffusion behavior, with CH_4 exhibiting a higher effective diffusivity [13]. Xiong et al. explored methane adsorption in organic-rich pores through experiments and molecular simulations. Their findings indicate that kerogen exhibits the highest methane adsorption capacity, followed by clay minerals and quartz. At a fixed pore size, the volume fraction of adsorbed gas decreases as pressure increases. Similarly, at a constant pressure, larger pores contain a lower proportion of adsorbed gas [14].

In reservoirs, besides gas adsorption behavior, water also plays an important role in gas production. Zhang et al. investigated supercritical methane adsorption in nanopores under initial water saturation conditions. They visualized water distribution within reservoir pore networks and analyzed its impact on methane adsorption at the microscopic level. Their results indicate that neglecting the influence of water distribution can lead to an overestimation of natural gas reserves [15]. Passey et al. demonstrated that the thickness of the water layer adsorbed on nanopore walls in the inorganic matrix is approximately equal to the pore diameter [16]. Shi et al. observed that the impact of the adsorbed water layer on the permeability of the inorganic matrix varies with water saturation [17]. Li et al. reported that an adsorbed water layer exists on clay mineral surfaces, with its thickness primarily determined by relative humidity and pore size [18]. Jin and Firoozabadi, using grand canonical Monte Carlo simulations, showed that adsorbed water layers form on clay mineral surfaces, restricting available space for gas flow [19]. Liu et al., through molecular dynamics simulations, found that water molecules exhibit stronger adsorption energy on clay mineral surfaces than methane molecules, leading to significant water molecule aggregation in the adsorption layer [20]. Xie et al. examined how water and salinity affect CO_2 adsorption and storage capacity in organic and clay nanopores. The results indicate that the presence of water reduces available adsorption space, leading to lower CO_2 storage capacity [21]. Zhang et al. studied H_2O and CH_4 flow in calcium montmorillonite nanoslits. They found that water forms bridges and thin films, blocking CH_4 flow at high water saturation (>80.87%) [22]. Zhang developed a kinetic model showing that water reduces CH_4 diffusion and adsorption while being more affected by pore structure. Water rapidly infiltrates pores, preventing CH_4 from reaching the surface [23].

In this study, molecular dynamics simulations were conducted to investigate the adsorption behavior and recovery degree evaluation of H_2O and CH_4 under high-temperature and high-pressure conditions. First, the wetting properties of the wall model were adjusted to match real reservoir conditions. Then, pore models for pure water and methane were established to analyze the effects of gas and water and varying water saturation on the density distribution under reservoir conditions. Subsequently, we conducted depletion development to investigate the recovery factor of pure gas and water under different water saturations. The results obtained from this study are significant for evaluating the development potential of CH_4 in ultra-deep tight sandstone gas reservoirs and provide certain reference values for enhancing the recovery factor of fractured tight sandstone gas reservoirs.

2. Methods

2.1. Model

The reservoir in the Keshen area of the Tarim Basin is mainly composed of quartz. Therefore, this study used the α-quartz single crystal cell to construct the wall model. As shown in Figure 1a, the lattice unit was obtained by cleaving the (1 0 0) surface (Figure 1b) and exposing the oxygen atoms, resulting in an orthogonal cell structure. Subsequently, the SiO_2 wall model was expanded, as shown in Figure 1c.

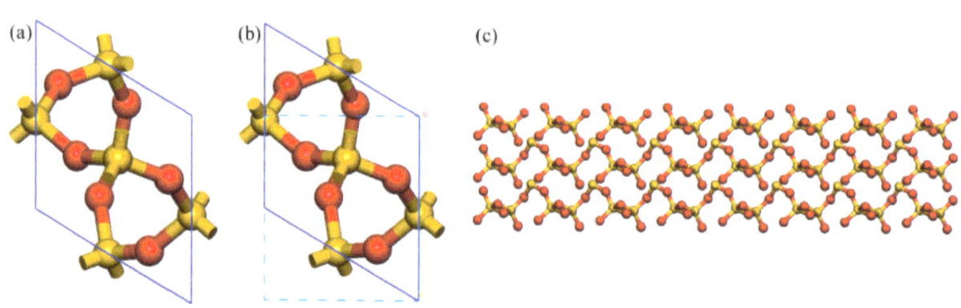

Figure 1. Construction of the quartz model: (**a**) crystal cell, (**b**) cleaved (1 0 0) surface, (**c**) nanopore model constructed.

Since the real reservoir conditions consist of dense sandstone with a minimum pore size of 10 nm, this study selected 10 nm as the representative pore size to construct a quartz wall model. This model was designed to investigate the effects of confinement on the development process. Based on this, a quartz wall with a 10 nm pore diameter was constructed using a modified wall structure. Additionally, graphene carbon plates were installed on both sides of the pore wall model, with one plate fixed and the other movable. By applying a specific acceleration to the movable carbon plate, different pressure conditions were simulated, providing the necessary experimental setup for subsequent research. As shown in Figure 2, the size of the entire model was 20.00 nm × 4.25 nm × 10.00 nm. In addition, the overall simulation system remained electrically neutral.

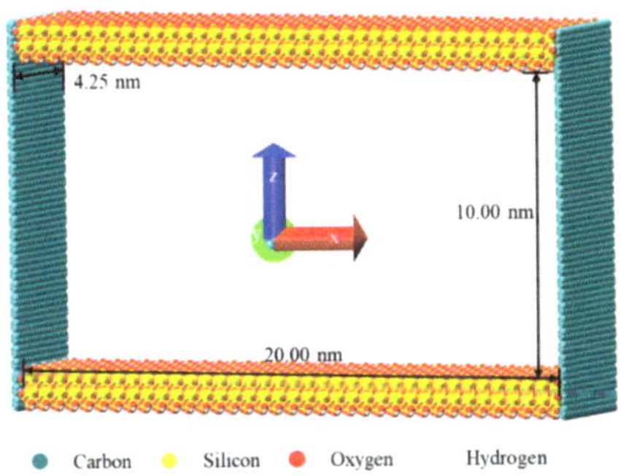

Figure 2. Quartz wall model with 10 nm pores.

2.2. Simulation Details

During the simulation, the quartz model was implemented using the CLAYFF force field [24]. To reduce the computational cost, a rigid wall was employed, meaning that wall atoms were fixed, and interactions between them were ignored. The water molecules were modeled using the SPC/E force field [25]. The CH_4 molecules were modeled using the TraPPE-UA force field [26]. The carbon atoms were modeled using the OPLS-AA force field [27]. The simulation was conducted using GROMACS software (version 2020.6) [28], with periodic boundary conditions applied in all three spatial directions (X, Y, Z). Molecular assembly was carried out using Packmol software (version 18.169) [29]. The NVT ensemble was used, maintaining a simulation temperature of 100 °C for 20 ns. After the simulation, VMD [30] was utilized to visualize molecular configurations and trajectories.

3. Result and Discussion

3.1. Wettability Adjustment

The reservoirs in the Keshen area of the Tarim Basin have a wettability angle of approximately 30°. To replicate this condition, the wall model was modified by introducing hydroxyl groups (–OH), which were generated by directly bonding hydrogen atoms to the exposed oxygen atoms. Quartz surface models with 0% (see Figure 3a) and 100% (see Figure 3b) hydroxylation were constructed. For the water droplet, a spherical water droplet with a density of 1.0 g/cm^3 was established and placed on the surface of the wall model to construct initial configurations. During the equilibration stage, water molecules from the final 5 ns of the simulation were selected to generate a 2D density cloud map.

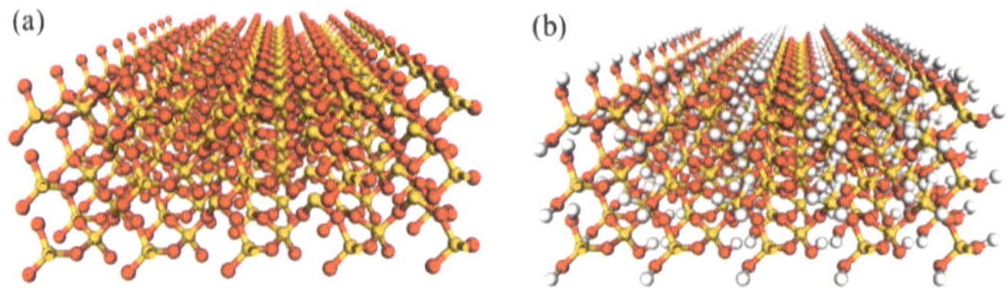

Figure 3. Wall models: (**a**) no –OH modification, (**b**) 100% –OH modification.

Figure 4a,c depicts the final configurations of the 0% and 100% hydroxylated models, respectively, while Figure 4b,d shows the corresponding 2D density cloud maps. The color gradient from red to blue represents a decrease in density. The wettability outline was extracted, and a tangent was drawn to determine the wettability angle. On the surface without hydroxyl modification, the water droplet is hemispherical after equilibrium, yielding a calculated wettability angle of 61.38°. As hydroxylation coverage increases, the droplet spreads more on the surface, enhancing wettability. When the surface is 100% hydroxylated, the wettability angle decreases to 29.77°, aligning with real reservoir conditions and consistent with literature reports [31].

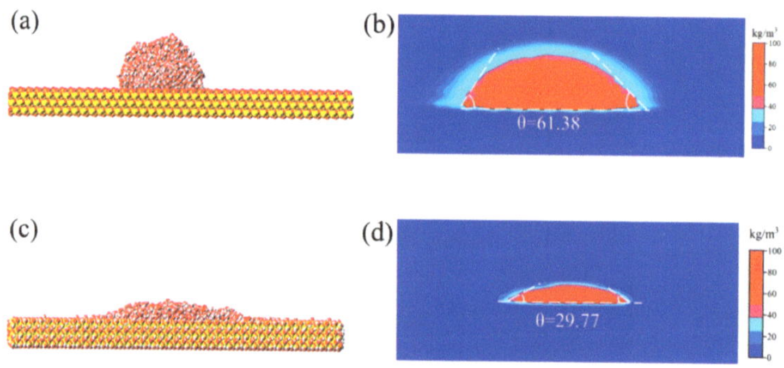

Figure 4. Balanced model configuration and 2D density cloud map. (**a**) Wettability configuration of the 0% –OH model; (**b**) density cloud map of the 0% –OH model; (**c**) wettability configuration of the 100% –OH model; (**d**) density cloud map of the 100% –OH model.

3.2. Density Distribution

To better understand the impact of adsorption on gas reservoir development, density distribution curves for each substance were recorded at 0.02 nm intervals after the simulation reached equilibrium. A detailed analysis of these curves allows for an intuitive observation of the adsorption behavior of gas and water on the wall and their variation patterns, providing insights into the role of adsorption in gas reservoir development assessment.

3.2.1. The Density Distribution of Pure H_2O

Figure 5 presents the density distribution curve of pure water under reservoir conditions (100 MPa, 100 °C). It can be seen from the figure that there is an adsorption zone of high density and a free zone of gentle density curve distribution in the quartz pores. The density of the adsorption layer is significantly higher than that of the free zone, indicating strong water molecule adsorption on the quartz pore wall under reservoir conditions. Specifically, water molecules form two high-density adsorption layers within 0.6 nm near the wall surface, with density peaks decreasing as the distance from the wall decreases. Each peak represents an adsorption layer, with the first adsorption layer closest to the wall reaching a density of 1.7 g/cm^3, approximately 1.7 times the bulk water density under the same temperature and pressure conditions. The second adsorption layer exhibits a lower density peak of 1.1 g/cm^3, showing that adsorption strength weakens as the distance from the wall increases. Beyond these adsorption layers, water density gradually decreases, and in the center of the pore, the density stabilizes at 0.99 g/cm^3, matching the bulk density. This suggests that the free zone behaves as a bulk phase. Xu et al. employed molecular dynamics simulations to investigate the flow behavior of alkane–water systems in quartz nanopores and found that under conditions of 50 °C and 20 MPa, water molecules formed four adsorption layers with a total thickness of 0.78 nm near the quartz surface. This result is in good agreement with the findings of the present study [32].

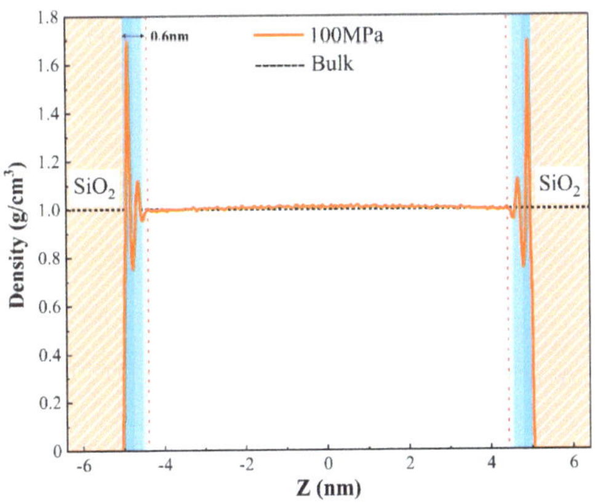

Figure 5. The density distribution curve of pure water at 100 MPa and 100 °C.

3.2.2. The Density Distribution of Pure CH_4

Figure 6 shows that the density distribution curve of pure CH_4 follows a similar trend to that of pure water, with regions of a high-density adsorption zone and regions where the density curve is gently distributed in the free zone, and the density of the adsorption layer is much higher than that of the fluid in the free zone. However, methane molecules form three adsorption layers on the quartz wall, with a greater adsorption layer thickness than that of water molecules, measuring 1.1 nm. This is because hydrogen bonding interactions cause water molecules to form a more compact molecular layer on the wall. Additionally, methane molecules have a larger molecular volume than water molecules, further contributing to the increased layer thickness. The first adsorption layer of methane, closest to the wall, has a density of approximately 0.42 g/cm^3, which is lower than that of water molecules (1.7 g/cm^3), confirming that water molecules near the wall surface are more compactly arranged. The second adsorption layer exhibits a lower density peak of 0.32 g/cm^3, while the density in the pore center stabilizes at 0.3 g/cm^3, consistent with the bulk density value. Ren et al. investigated methane adsorption in 10 nm quartz pores under conditions of 110 °C and 30 MPa and reported an adsorption layer thickness of approximately 1 nm, which is in good agreement with the present findings [12].

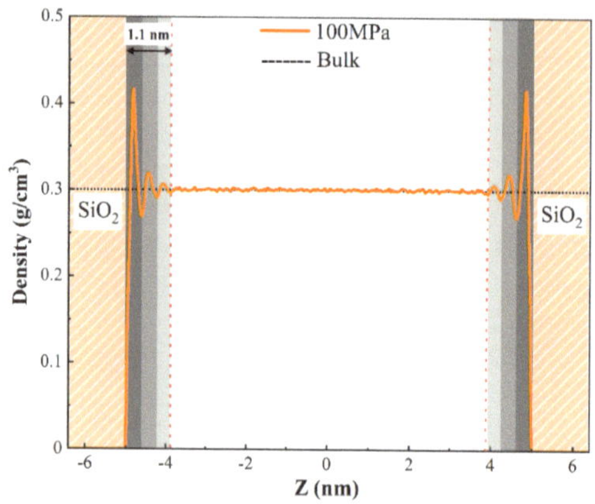

Figure 6. Density distribution curve of pure CH_4 at 100 MPa and 100 °C.

3.2.3. The Density Distribution of Gas-Water Under Different Water Saturation Conditions

Under gas–water coexistence conditions, due to the hydrophilicity of the quartz wall, H_2O and CH_4 compete for adsorption, with water molecules preferentially occupying specific positions on the wall surface. As shown in the final molecular configuration of Figure 7a, at 30% water saturation, water molecules form a thin water film along the two sides of the wall surface. Consequently, no CH_4 adsorption layer forms near the wall, and CH_4 exists solely in the pore center as a free phase. The gas phase exists as a continuous phase in the pore center. At this point, the CH_4 density in the pore center is approximately 0.3 g/cm^3.

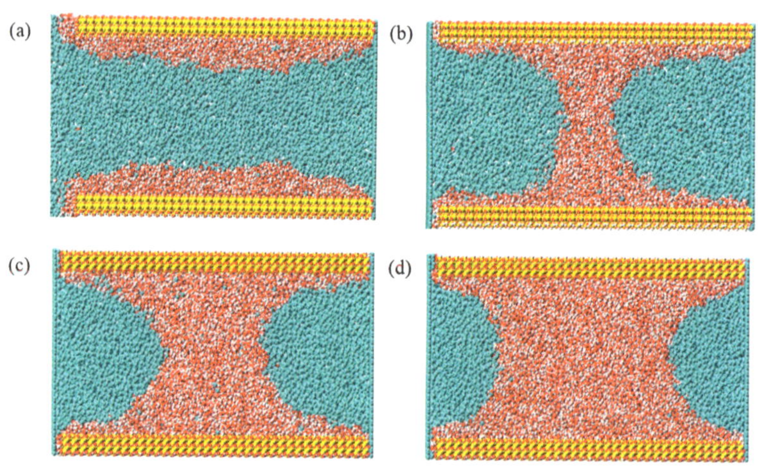

Figure 7. Distribution of H_2O/CH_4 under different water saturation conditions. (**a**) The configuration at 30% water saturation, (**b**) the configuration at 35% water saturation, (**c**) the configuration at 50% water saturation, and (**d**) the configuration at 70% water saturation.

As water saturation increases, water molecules aggregate under the action of hydrogen bonds, and at 35% water saturation, liquid bridges are formed in the pores, at which time the gas phase is divided into different regions, forming discrete bubbles and trapped gases (see Figure 7b). The gas–water density distribution curve in Figure 8b shows that the water density in the pore center increases to 0.13 g/cm^3, while the gas density decreases to 0.23 g/cm^3. With further increases in water saturation, the thickness of the liquid bridge thickens (see Figure 7c,d), intensifying the gas-trapping effect. When water saturation reaches 70%, the final molecular configuration reveals a significant increase in water volume within the pore, drastically reducing the available space for CH_4. Additionally, the water density in the free zone also increases (see Figure 8d). At this stage, the water density in the pore center reaches 0.50 g/cm^3, while the CH_4 density decreases to 0.14 g/cm^3.

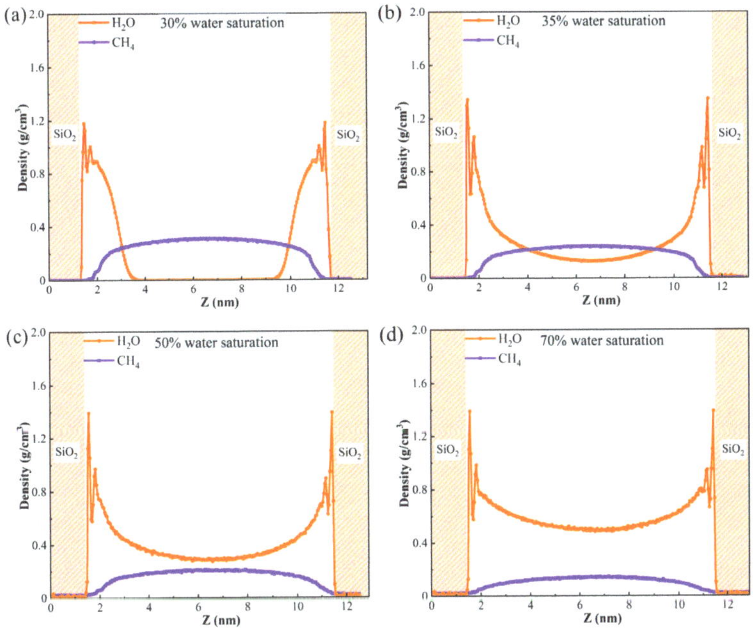

Figure 8. Density distribution of H_2O/CH_4 under different water saturation conditions. (**a**) Density distribution with a 30% water saturation, (**b**) density distribution with a 35% water saturation, (**c**) density distribution with a 50% water saturation, and (**d**) density distribution with a 70% water saturation.

3.3. Mobility Evaluation

To better understand the adsorption effect of the wall and the influence of water on the recovery factor, this study evaluated the recovery degree of the adsorption layer and the free zone during the depletion development and discussed the impact of different water saturations on the recovery factor of the gas and water. In the simulation, the depletion development process was achieved in the following way. The final molecular configuration of the gas–water distribution state under reservoir pressure was taken as the initial configuration for the depletion development simulation. Then, according to the requirements of depletion to different pressures, the acceleration applied to the carbon plate was changed, thereby realizing the depletion process from 100 MPa to 10 MPa. During each depletion development, after the system was completely balanced, the last 5 ns were taken to analyze the distribution state of the gas and water and the corresponding recovery factor.

3.3.1. Mobility Evaluation of Pure H_2O

As illustrated in Figure 9a, the water molecular system exhibits low elastic potential energy, indicating minimal volumetric change even as pressure decreases from 100 MPa to 10 MPa. This suggests that pure water remains largely immobile during depletion development due to its limited compressibility. At the microscopic level, strong hydrogen bonding and intermolecular interactions stabilize the water structure, restricting large-scale molecular rearrangement and migration. Consequently, water in the adsorbed zone remains strongly bound to the pore walls, exhibiting extremely low mobility, whereas water in the free zone, though not directly influenced by surface adsorption, is still constrained by intermolecular forces, leading to similarly limited displacement. Figure 9b further reveals that the recovery factors in both zones remain low (<10%), with 4% in the adsorbed zone and 8% in the free zone, indicating that free-phase water demonstrates relatively higher sensitivity to pressure depletion. This can be attributed to the absence of direct solid-phase interactions, allowing for greater molecular displacement compared to the adsorbed phase. These findings highlight a fundamental challenge in ultra-deep fractured tight reservoirs, where water mobility is highly restricted under the depletion process, potentially hindering gas flow and reducing the ultimate recovery factor.

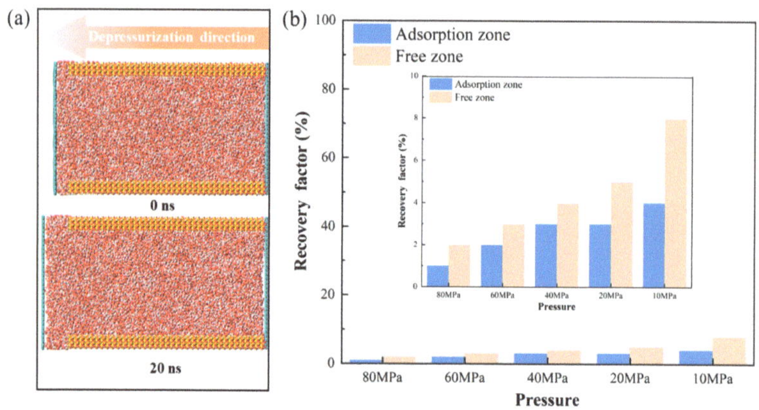

Figure 9. Adsorption and recovery factor of pure H_2O under different pressures: (**a**) the configurations of pure H_2O in nanopores, (**b**) the recovery factor in the different zones.

3.3.2. Mobility Evaluation of Pure CH_4

Figure 10a illustrates the molecular configurations of pure CH_4 during the depletion development process. It is visually evident that as pressure decreases, the gas phase undergoes significant expansion, further confirming the high sensitivity of CH_4 to pressure changes, which is notably different from the flow behavior of pure water. Specifically, when pressure decreases from 100 MPa to 80 MPa, the recovery factors of the adsorption layer and the free zone are 9% and 10%, respectively, both significantly higher than those of water, indicating that gas desorption and diffusion begin even at high pressures. As the pressure further decreases to 10 MPa, the recovery factors of the adsorption layer and the free zone increase substantially. This result suggests that during the depletion development process, pure CH_4 exhibits strong mobility, with gas in the free zone being more readily mobilized than that in the adsorption layer. Figure 10b further presents the recovery factor trends of CH_4 in the adsorption and free zones under different pressures. As pressure decreases, gas in all regions of the pore is progressively mobilized, and the recovery factor of the free zone consistently exceeds that of the adsorption layer. When the pressure ultimately drops to 10 MPa, the recovery factor of the adsorption layer reaches 56%, whereas the recovery factor of the free zone reaches 79%, indicating that free-phase gas has a stronger release capacity during pressure depletion. This phenomenon can be attributed to the fact that free-zone gas is not constrained by solid surface adsorption, allowing its molecules to diffuse and migrate more easily. In contrast, gas in the adsorption layer remains influenced by van der Waals forces and adsorption energy from the solid surface, leading to a lower desorption rate compared to the free zone. Nevertheless, the final recovery factor of the adsorption zone remains relatively high, demonstrating that gas desorption plays a significant role in the depletion-driven development process, highlighting the importance of pressure management strategies in enhancing ultimate gas recovery in ultra-deep fractured tight reservoirs.

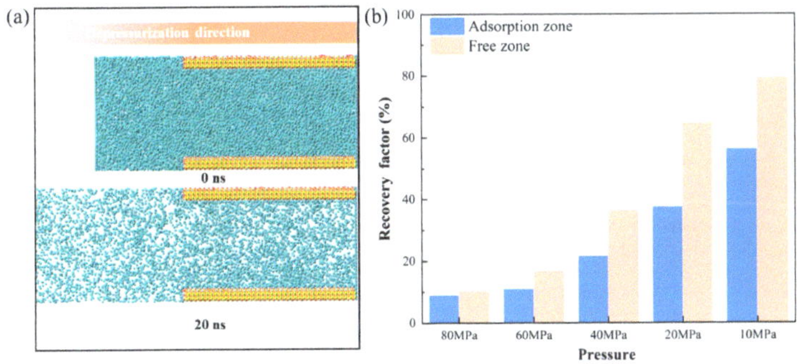

Figure 10. Adsorption and recovery factor of pure CH_4 under different pressures: (**a**) the configurations of pure CH_4 in nanopores, (**b**) the recovery factor in the different zones.

Figure 11 illustrates the overall recovery factors of pure H_2O and pure CH_4 under depletion development. The results indicate that while the recovery factor of pure water increases slightly with decreasing pressure, the overall increase remains minimal, with a final recovery factor of only 7%. This suggests that water exhibits low mobility under pressure depletion, likely due to its strong intermolecular hydrogen bonding and interaction with the pore walls, which restrict its movement even under significant pressure reduction. In contrast, the recovery factor of pure CH_4 shows a dramatic increase as pressure decreases, ultimately reaching 75%. This significant difference highlights the much higher compressibility and mobility of CH_4 compared to water, allowing gas molecules to expand and migrate more effectively during the depletion process. The stark contrast between H_2O and CH_4 mobility further confirms that gas production in ultra-deep reservoirs is highly pressure-sensitive, whereas water remains largely immobile, leading to potential challenges in managing residual water saturation and optimizing overall recovery efficiency.

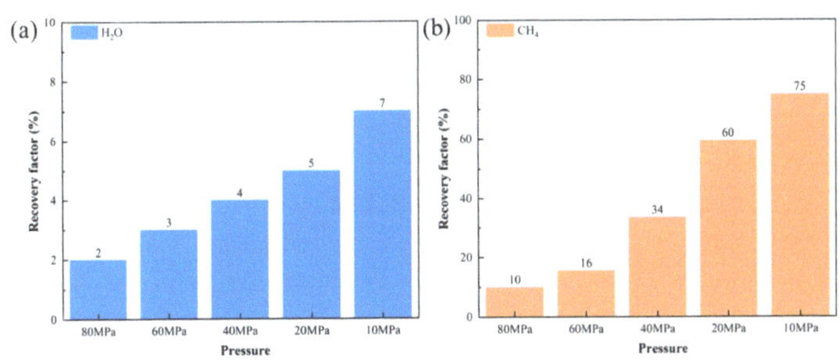

Figure 11. Recovery factor of H_2O/CH_4 under different pressures: (**a**) recovery factor of H_2O under different pressures, (**b**) recovery factor of CH_4 under different pressures.

3.3.3. Mobility Evaluation of Gas-Water Under Different Water Saturation Conditions

In real reservoirs, water is commonly present. To investigate the recovery degree of gas and water in the pore, the density distribution and the recovery factor under different pressures were studied at water saturation of 30%, 35%, 50%, and 70%. The following is a specific analysis.

At 30% water saturation, there is a significant difference in the adsorption behavior of the gas and water phases within the pore. As shown in Figure 12a, the final molecular configuration indicates that due to the hydrophilicity of the pore wall, H_2O and CH_4 undergo competitive adsorption, and water molecules primarily adsorb onto the pore wall, forming a stable water film that completely covers the wall surface, causing the gas phase to exist only as a continuous phase in the pore center. Due to the strong wall–fluid interactions, the water film remains firmly adsorbed and exhibits minimal mobility, resulting in an extremely low final recovery factor of only 1%, indicating that the water film is hardly mobilized during pressure depletion. Water molecules form a continuous thin water film along the pore walls due to the hydrophilicity of quartz. As a result, CH_4 is excluded from the adsorption zone and exists entirely as a free phase in the pore center. Compared to the dry condition (pure CH_4), where CH_4 is partially adsorbed on the wall and more strongly confined, the CH_4 in the free zone under 30% water saturation is more mobile and responds more sensitively to pressure depletion. This structural change leads to a slightly higher recovery factor. It leads to a final recovery factor of 82% (Figure 12b), even exceeding the overall recovery factor of pure CH_4 (75%). Furthermore, these findings suggest that in the presence of water, the existence of the water phase may alter the gas distribution state, causing CH_4 to exist primarily as a free phase, thereby further enhancing its recoverability.

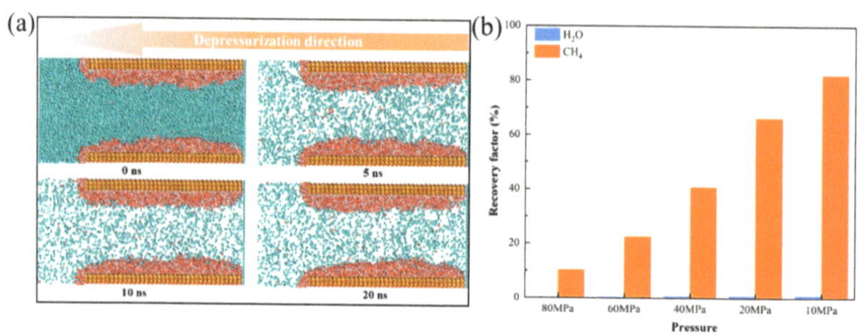

Figure 12. 30% water saturation: (**a**) snapshots at different times, (**b**) recovery factor at different pressures.

As the water saturation increases to 35%, water molecules within the nanopores aggregate through hydrogen bonding, forming liquid bridge structures (see Figure 13a, 0 ns). At this stage, the gas phase becomes segmented into different isolated regions, leading to the formation of trapped gas and gas bubbles. The presence of liquid bridges enhances the connectivity of the water phase within the pore while reducing the continuity of the gas phase, thereby restricting gas flow. With the decrease in pressure, the gas expands and flows out, gradually breaking through the liquid bridges and causing the water phase structure to rearrange. The liquid bridges progressively rupture and redistribute near the pore walls, forming thin liquid films (see Figure 13a, 20 ns). Under this water saturation condition, the gas-driven migration of liquid bridges during pressure depletion enhances water mobilization, leading to an increased final recovery factor of 8%, which represents an improvement compared to lower water saturation conditions. However, despite the increased water mobilization, gas mobility is adversely affected. As gas flows through the pore, it must overcome the blocking effect of the liquid film on the flow pathways, resulting in a decrease in overall gas recovery, which ultimately drops to 78% (see Figure 13b), lower than the gas recovery observed at 30% water saturation. These findings indicate that under moderate water saturation conditions, the formation and rupture of liquid bridges play a crucial role in the gas–water two-phase flow. While water mobilization is enhanced due to liquid bridge migration driven by pressure depletion, the obstruction effect of liquid bridges on gas flow pathways reduces the effective gas recovery. Therefore, in the development of high water saturation reservoirs, optimizing the pressure depletion rate to enhance water mobilization while minimizing gas flow resistance is a key strategy for improving recovery efficiency.

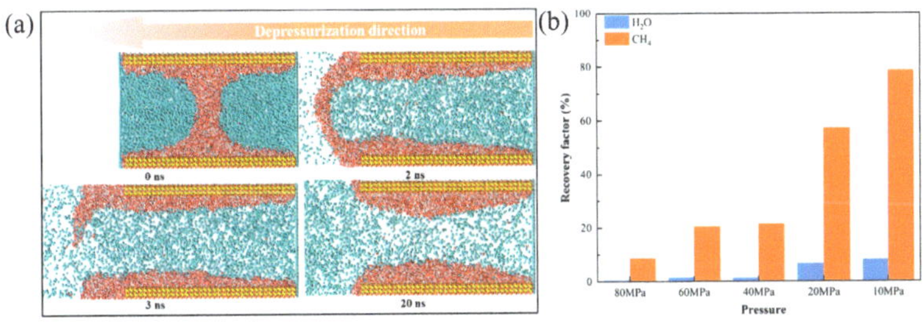

Figure 13. 35% water saturation: (**a**) snapshots at different times, (**b**) recovery factor at different pressures.

At 50% water saturation, it can be observed that as water saturation increases, the distribution of the water phase within the pore undergoes significant changes, with the thickness of liquid bridges increasing, leading to the occurrence of microscopic water blocking (Figure 14a, 5 ns). During the depletion process, although the expansion of gas drives some movement of the water phase, the liquid bridges remain intact and do not rupture. Meanwhile, the trapped gas phase enclosed by the liquid bridges fails to break through the water phase barriers, restricting gas flow pathways (Figure 14a, 15 ns). Figure 14b further reveals the impact of water saturation on the gas–water two-phase flow and mobilization. Due to the increased thickness of the liquid bridges, the expansion of gas enhances water mobilization, resulting in a final water recovery factor of 57%, which represents a 49% increase compared to the 35% water saturation condition. However, despite the improvement in water mobilization, the mobility of CH_4 is significantly restricted, leading to a final CH_4 recovery factor of only 39%. This indicates that as water saturation further increases, the blocking effect of the water phase within the pore intensifies, causing a substantial decline in the gas recovery factor.

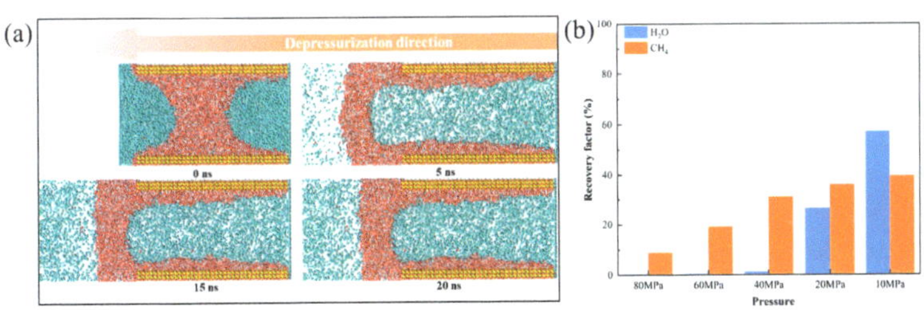

Figure 14. 50% water saturation: (**a**) snapshots at different times, (**b**) recovery factor at different pressures.

When the water saturation increases to 70%, water molecules occupy the vast majority of the pore space, leaving only a small amount of gas trapped in the remaining pores, as shown in Figure 15a. At this point, the liquid bridges thicken, forming a more stable blocking structure that further restricts gas flow pathways. In the subsequent decompression process, the influence of gas on liquid bridge flooding was significantly weakened due to the increase in liquid bridge thickness and the decrease in gas content. As a result, the recovery factor of the water phase decreases, ultimately reaching only 46%. Furthermore, in a high water saturation environment, the strong blocking effect of liquid bridges further limits the mobilization of gas, leading to a significant reduction in the final CH_4 recovery factor, which drops to only 29% (Figure 15b). Compared to the pure gas condition, the gas recovery factor is reduced by 46%. This phenomenon indicates that under high water saturation conditions, the water phase not only occupies a large portion of the pore space, but also severely inhibits CH_4 percolation and production through enhanced blocking effects. The stability of liquid bridges and their obstructive influence on gas flow pathways have become critical factors affecting the efficiency of gas reservoir development.

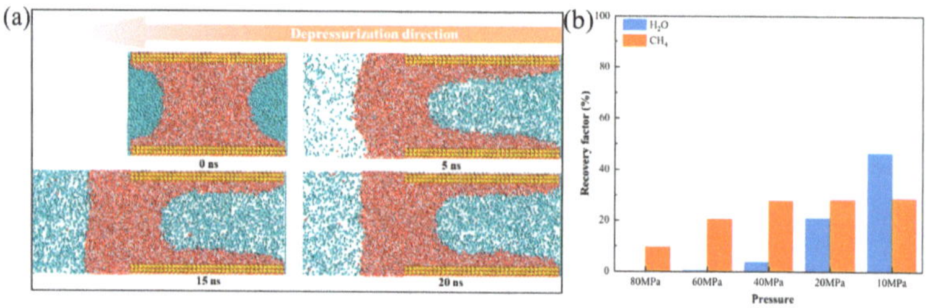

Figure 15. 70% water saturation: (**a**) snapshots at different times, (**b**) recovery factor at different pressures.

4. Conclusions

This study employed molecular simulation to systematically analyze the density distribution, adsorption behavior, and mobilization characteristics of gas and water in 10 nm pores of the Keshen tight sandstone gas reservoir in the Tarim Basin. Additionally, it evaluated the evolution of gas and water mobilization during depletion development. The results indicate that H_2O and CH_4 exhibit distinct adsorption behaviors. Near the quartz wall, H_2O forms two adsorption layers with a total thickness of 0.6 nm, whereas CH_4 forms three adsorption layers with a total thickness of 1.1 nm. During depletion, H_2O is constrained by strong wall adsorption, leading to low mobilization, with recovery factors of only 4% and 8% in the adsorption and free zones, respectively. In contrast, CH_4 is highly sensitive to pressure changes, achieving recovery factors of 56% and 79% in the adsorption and free zones, respectively. These findings indicate that CH_4 exhibits greater mobility under pressure depletion, while H_2O remains largely immobile due to strong wall adsorption.

Under water-bearing conditions, the presence of water significantly affects the distribution and mobilization of CH_4. Due to the hydrophilic nature of the quartz wall, H_2O preferentially adsorbs onto the wall surface, dominating the adsorption process and significantly inhibiting CH_4 adsorption. At 30% water saturation, CH_4 mainly exists in the pore center as a free phase, maintaining relatively high mobility. However, as water saturation increases, CH_4 mobilization becomes significantly restricted. When water saturation increases from 30% to 70%, the water phase forms stable liquid bridges within the pores, segregating the gas phase into isolated regions. As water saturation further increases, microscopic water lock effects emerge, further inhibiting CH_4 flow, and the recovery factor of CH_4 decreases significantly from 75% to 29%.

This study provides a microscale perspective on gas–water transport mechanisms in ultra-deep tight sandstone gas reservoirs and reveals the impact of water phase blocking on CH_4 mobility, which is crucial for optimizing gas extraction strategies in high-water-saturation reservoirs. However, this study is based on molecular simulations of 10 nm pores, without considering complex pore networks or multiscale effects. Future studies should incorporate larger-scale pore network modeling, high-pressure and high-temperature experimental validation, and coupled flow simulations to further improve the understanding of gas–water transport mechanisms in ultra-deep tight sandstone reservoirs.

Author Contributions: Conceptualization, Y.L. (Yongfu Liu); data curation, X.P., C.W. and Y.L. (Yijia Li); formal analysis, F.Y. and L.D.; investigation, Y.L. (Yongfu Liu) and X.P.; methodology, X.P.; resources, F.Y., L.D., T.Z. and S.X.; software, X.P.; supervision, J.Z., Y.L. (Yijia Li) and S.X.; validation, T.Z.; visualization, C.W.; writing—original draft preparation, Y.L. (Yongfu Liu); writing—review and editing, Y.L. (Yongfu Liu) and J.Z. All authors have read and agreed to the published version of the manuscript.

Funding: This research received no external funding.

Data Availability Statement: The original contributions presented in the study are included in the article, further inquiries can be directed to the corresponding author.

Conflicts of Interest: Authors Yongfu Liu, Xuehao Pei, Fenglai Yang, Li Dai, Cuili Wang, Tingya Zhou, Yijia Li and Sa Xiao were employed by the company PetroChina. Authors Yongfu Liu, Fenglai Yang and Cuili Wang were employed by the company China National Petroleum Corporation. The remaining authors declare that the research was conducted in the absence of any commercial or financial relationships that could be construed as a potential conflict of interest.

References

1. He, D.; Jia, C.; Zhao, W.; Xu, F.; Luo, X.; Liu, W.; Tang, Y.; Gao, S.; Zheng, X.; Li, D.; et al. Research progress and key issues of ultra-deep oil and gas exploration in China. *Pet. Explor. Dev.* **2023**, *50*, 1333–1344. [CrossRef]
2. Guo, X.; Zhao, L.; Han, W.; Zhou, L.; Huang, Z.; Sun, X.; Yang, X.; Zhang, T.; Zhang, C. Geochemistry of Formation Water and Implications for Ultradeep Tight Sandstone of DK Gas Field in Kuqa Depression. *Geofluids* **2022**, *2022*, 6514733. [CrossRef]
3. Lai, J.; Wang, G.; Chai, Y.; Xin, Y.; Wu, Q.; Zhang, X.; Sun, Y. Deep burial diagenesis and reservoir quality evolution of high-temperature, high-pressure sandstones: Examples from Lower Cretaceous Bashijiqike Formation in Keshen area, Kuqa depression, Tarim basin of China. *AAPG Bull.* **2017**, *101*, 829–862. [CrossRef]
4. Zhao, L.B.; Yang, X.J.; Chang, L.J.; Zhang, T.H. Tri-media reservoir characteristic of fractured tight gas reservoir of gasfield A, Kuqa Depression, Tarim Basin. *Nat. Gas Geosci.* **2017**, *28*, 209–218. [CrossRef]
5. Jiang, T.; Sun, X. Development of Keshen ultra-deep and ultra-high pressure gas reservoirs in the kuqa foreland basin, Tarim Basin: Understanding points and technical countermeasures. *Nat. Gas Ind.* **2018**, *6*, 16–24. [CrossRef]
6. Wang, R.; Zhang, C.; Chen, D.; Yang, F.; Li, H.; Li, M. Microscopic Seepage Mechanism of Gas and Water in Ultra-Deep Fractured Sandstone Gas Reservoirs of Low Porosity: A Case Study of Keshen Gas Field in Kuqa Depression of Tarim Basin, China. *Front. Earth Sci.* **2022**, *10*, 893701. [CrossRef]
7. He, X.; Wang, C.; Chang, B.; Cao, Z.; Tang, H. A Dynamic Reserve Evaluation Method for an Ultra-Deep Fractured Tight Sandstone Gas Reservoir. *Energies* **2024**, *17*, 2648. [CrossRef]
8. Lei, Q.; Xu, Y.; Yang, Z.; Cai, B.; Wang, X.; Zhou, L.; Liu, H.; Xu, M.; Wang, L.; Li, S. Progress and development directions of stimulation techniques for ultra-deep oil and gas reservoirs. *Pet. Explor. Dev.* **2021**, *48*, 221–231. [CrossRef]
9. Li, Q.; Li, Q.; Wu, J.; Li, X.; Li, H.; Cheng, Y. Wellhead Stability During Development Process of Hydrate Reservoir in the Northern South China Sea: Evolution and Mechanism. *Processes* **2025**, *13*, 40. [CrossRef]
10. Li, Q.; Li, Q.; Cao, H.; Wu, J.; Wang, F.; Wang, Y. The Crack Propagation Behaviour of CO_2 Fracturing Fluid in Unconventional Low Permeability Reservoirs: Factor Analysis and Mechanism Revelation. *Processes* **2025**, *13*, 159. [CrossRef]
11. Zhang, R.; Tang, Y. Molecular dynamics simulation on the density distribution and multilayer adsorption of methane in nanopores. *Phys. Fluids* **2024**, *36*, 122001. [CrossRef]
12. Ren, J.H.; Ren, X.H.; Song, H.Q.; Han, D.L.; Wang, C.C.; Sheng, G.L.; Lv, F.W. Adsorption and diffusion characteristics of methane in nanopores based on molecular simulation. *Acta Pet. Sin.* **2020**, *41*, 1366–1375. [CrossRef]
13. Zhang, H. Nanoscale Transport of Multicomponent Fluids in Shales. Ph.D. Thesis, Virginia Polytechnic Institute and State University, Blacksburg, VA, USA, 2025.
14. Xiong, J.; Liu, X.; Liang, L.; Zeng, Q. Adsorption of methane in organic-rich shale nanopores: An experimental and molecular simulation study. *Fuel* **2017**, *200*, 299–315. [CrossRef]
15. Zhang, L.; Zhang, T.; Zhao, Y. Methane Adsorption in Nanoporous Shale. In *Modelling in Nanoporous Shale*; Advances in Oil and Gas Exploration & Production; Springer: Cham, Switzerland, 2024; pp. 71–83. [CrossRef]

16. Passey, Q.R.; Bohacs, K.M.; Esch, W.L.; Klimentidis, R.; Sinha, S. From oil-prone source rock to gas-producing shale reservoir–geologic and petrophysical characterization of unconventional shale-gas reservoirs. In Proceedings of the SPE International Oil and Gas Conference and Exhibition in China, Beijing, China, 8–10 June 2010; SPE: Richardson, TX, USA, 2010.
17. Shi, J.; Zhang, L.; Li, Y.; Yu, W.; He, X.; Liu, N.; Li, X.; Wang, T. Diffusion and Flow Mechanisms of Shale Gas through Matrix Pores and Gas Production Forecasting. In Proceedings of the SPE Unconventional Resources Conference Canada, Calgary, AB, Canada, 5–7 November 2013; OnePetro: Richardson, TX, USA, 2013. [CrossRef]
18. Li, J.; Li, X.; Wang, X.; Li, Y.; Wu, K.; Shi, J.; Yang, L.; Feng, D.; Zhang, T.; Yu, P. Water distribution characteristic and effect on methane adsorption capacity in shale clay. *Int. J. Coal Geol.* **2016**, *159*, 135–154. [CrossRef]
19. Jin, Z.; Firoozabadi, A. Effect of water on methane and carbon dioxide sorption in clay minerals by Monte Carlo simulations. *Fluid Phase Equilibria* **2014**, *382*, 10–20. [CrossRef]
20. Liu, J.; Wen, Y.; Jiang, L.; Yuan, H.; Liu, Z.; Shui, Z.; Wang, B.; Yan, X. Molecular dynamics simulation of micro mechanisms of gas and water migration and diffusion in shale reservoirs. *Colloids Surf. Physicochem. Eng. Asp.* **2024**, *682*, 132865. [CrossRef]
21. Xie, C.; Huang, J.; Jiang, S.; Zhao, H.; Wu, Z. Effect of Water Content and Salinity on CH4/CO2 Competitive Adsorption in Organic and Clay Nanopores: A Molecular Perspective. *Energy Fuels* **2024**, *38*, 23507–23518. [CrossRef]
22. Zhang, L.; Yan, W.; Fu, J.; Cai, S.; Liang, H. Methane Gas Transport in Ca-MMT Shale Nanoslits Considering Water Content Effects: Insights from Molecular Dynamics Simulations. *Langmuir* **2024**, *40*, 25110–25117. [CrossRef]
23. Zhang, Z.; Qiu, M.; Zhang, D. Analytical simulation of the simultaneous adsorption process of methane and water vapor in shales. *Geosystem Eng.* **2025**, *28*, 31–45. [CrossRef]
24. Cygan, R.T.; Liang, J.-J.; Kalinichev, A.G. Molecular Models of Hydroxide, Oxyhydroxide, and Clay Phases and the Development of a General Force Field. *J. Phys. Chem. B* **2004**, *108*, 1255–1266. [CrossRef]
25. Berendsen, H.J.C.; Grigera, J.R.; Straatsma, T.P. The missing term in effective pair potentials. *J. Phys. Chem.* **1987**, *91*, 6269–6271. [CrossRef]
26. Eggimann, B.L.; Sunnarborg, A.J.; Stern, H.D.; Bliss, A.P.; Siepmann, J.I. An Online Parameter and Property Database for the TraPPE Force Field. *Mol. Simul.* **2014**, *40*, 101–105. [CrossRef]
27. Jorgensen, W.L.; Maxwell, D.S.; Tirado-Rives, J. Development and Testing of the OPLS All-Atom Force Field on Conformational Energetics and Properties of Organic Liquids. *J. Am. Chem. Soc.* **1996**, *118*, 11225–11236. [CrossRef]
28. Abraham, M.J.; Murtola, T.; Schulz, R.; Páll, S.; Smith, J.C.; Hess, B.; Lindahl, E. GROMACS: High performance molecular simulations through multi-level parallelism from laptops to supercomputers. *SoftwareX* **2015**, *1*, 19–25. [CrossRef]
29. Martínez, L.; Andrade, R.; Birgin, E.G.; Martínez, J.M. PACKMOL: A package for building initial configurations for molecular dynamics simulations. *J. Comput. Chem.* **2009**, *30*, 2157–2164. [CrossRef]
30. Humphrey, W.; Dalke, A.; Schulten, K. VMD: Visual molecular dynamics. *J. Mol. Graph.* **1996**, *14*, 33–38. [CrossRef]
31. Wang, Z.; Yu, C.; Zhao, J.; Guo, P.; Liu, H. Molecular dynamics simulation for quantitative characterization of wettability transition on silica surface. *J. Mater. Res. Technol.* **2022**, *19*, 4371–4380. [CrossRef]
32. Xu, J.; Zhan, S.; Wang, W.; Su, Y.; Wang, H. Molecular dynamics simulations of two-phase flow of n-alkanes with water in quartz nanopores. *Chem. Eng. J.* **2022**, *430*, 132800. [CrossRef]

energies

MDPI

Article

Applying Deep Electrical-Resistivity Tomography Techniques for the Exploration of Medium- and Low-Geothermal Energy Resources

Cristina Sáez Blázquez *, Ignacio Martín Nieto, Javier Carrasco, Pedro Carrasco, Daniel Porras, Miguel Ángel Maté-González, Arturo Farfán Martín and Diego González-Aguilera

Department of Cartographic and Land Engineering, Higher Polytechnic School of Avila, University of Salamanca, Hornos Caleros 50, 05003 Avila, Spain; nachomartin@usal.es (I.M.N.); tgeofisicas@gmail.com (J.C.); retep81@usal.es (P.C.); dporras@geoland.es (D.P.); mategonzalez@usal.es (M.Á.M.-G.); afarfan@usal.es (A.F.M.); daguilera@usal.es (D.G.-A.)

* Correspondence: u107596@usal.es

Abstract: The growth of the geothermal industry demands the constant search of techniques with the aim of reducing exploration efforts whilst minimizing subsurface uncertainty. The exploration of geothermal resources is fundamental from the exploitation point of view, especially in those regions where this energy is not as widespread as the rest of renewable sources. This research shows how geoelectrical methods can contribute to the investigation and characterization of medium–low enthalpy geothermal resources until about 800 m of depth. A 2000 m long electrical-resistivity tomography profile was performed in a region of Southern Spain with previous evidence of moderate geothermal potential. Results of this geophysical campaign (together with a preliminary geological characterization) allowed for the obtainment of a 2D profile and a pseudo-3D model with extensive information about the subsoil in terms of geological composition and formations. The interpretation of geophysical results denotes the existence of a potential formation constituted by carbonate materials with thickness greater than 300 m, crossing different fractures. Once the ideal location for the geothermal exploitation is defined, the research evaluates the contribution of the possible energy source, deducing that the energy extraction in the potential fracturing area would be double that of the one in the vicinity of the site.

Keywords: geothermal resources; exploration; electrical resistivity tomography; 2D profile; potential area

1. Introduction

Due to the current energy context and the exponential increase of the world energy demand, there is a clear need to move towards the massive use of renewable energy resources and become less dependent on social or geopolitical factors. In fact, the continuous availability of energy is considered one of the most critical aspects for society development, especially considering that, today, a significant portion of energy still derives from burning limited organic fuels. One approach to replace conventional fossil fuels and alleviate the mentioned energy issues is the introduction of renewable energy technologies. In this sense, geothermal energy appears as a potential contributor in the way of reducing the external and internal energy dependence on non-renewable systems [1], constituting one of the most efficient sources that can operate continuously to meet the energy demand 24/7 [2]. This energy can be directly used for heating and/or cooling applications, constituting one of the oldest and most versatile ways of utilizing geothermal resources [3]. In this context, the globally installed capacity for direct geothermal use worldwide was, at the end of 2019, 107,727 MWt, meaning a 52% increase over the year 2015, which is a growing annual rate of 8.7% [4]. The reason for this significant growth is, mainly, the recent technological development of geothermal heat pumps (GHPs), which generally constitute these shallow systems through the well-known ground-source or ground-water heat pump (GSHP-GWHP) technologies [5,6]. The alternative use of geothermal energy is power generation, which is usually attributed to deep and hydrothermal resources with an average growth rate of around 5% per year and a global production of 95 TWh in 2020 in more than 30 countries [7].

Despite the clear versatility of geothermal resources, when compared with other renewable solutions (e.g., biomass, hydro, solar PV, wind), geothermal falls far behind both in production and installed capacity. The principal obstacle to geothermal growth is the initial investment costs generally associated with power projects, but also for domestic heating and cooling solutions. However, numerous countries have done the required groundwork to conduct resource inventories with the aim of quantifying their potential for exploiting the different possibilities of geothermal energy [8,9].

In the particular case of Spain (where this research is focused), geothermal development is still undervalued. There are no high-enthalpy geothermal facilities operating in the country, and the shallow geothermal use is also limited. In this sense, although slower than desirable considering the existing capacity and energy requirements, the installation of GHP systems has been a growing trend in the last few years. In addition, public administrations are making a great effort to introduce GHP systems in public buildings (both in refitted and new spaces) in order to move into the concept of "Nearly Zero-Energy Buildings" (NZEBs) promoted by the European Union [10–13]. Based on the data from the World Geothermal Congresses of 2015 and 2020, the estimated uses for the country are summarized in the following Table 1.

Table 1. Estimated shallow geothermal uses for Spain [14,15].

Use	MWt	TJ/Year
Individual space heating	5.20	133.6
Greenhouse heating	22.0	165.4
Bath and swimming	3.80	92.0
Geothermal heat pumps	513.0	3542.0
Total for the country	**544.0**	**3933.0**

In spite of the clear advantages and the proven capacity to provide energy at a constant pace, as shown in the previous Table 1, the widespread production of geothermal energy in the country considered here has been limited by different factors. Among them are the lack of access to thermal supplies, the operating risks when drilling the geothermal wells, and the associated high capital costs [16,17]. When these factors are analyzed, it is common to conclude that most of these risks are mainly due to the lack of knowledge and precise characterization of the ground where the system is planned. In this sense, the in-depth evaluation of the subsoil structure and the determination of its expected thermal behavior are essential when designing the geothermal well field.

Regarding the reservoir temperature, it can be measured directly by bottom-hole temperature measurements. However, well measurements may not be representative of the entire reservoir, and geophysical prospecting appears as an effective tool for providing a more spatially complete information source. Depending on the type of geothermal system considered, the objective of these tests may be the determination of the geological structures and their distribution in depth, the estimation of the thermal properties of the existing formations, or the location of aquifers or singular structures, among many other applications [18–20]. The implementation of these techniques has proven to be useful when designing a shallow geothermal system and, thus, ensuring its correct operation during the estimated useful life period. But these prospecting systems are also essential when trying to achieve a better understanding of a deep geothermal resource. Geophysical detection and the monitoring of deep reservoirs represent a great advance in the exploration of geothermal energy and, ultimately, in the acquisition of complete and optimized information about the structure in depth and the possibilities of future geothermal exploitation [21,22].

In the context of defining new possibilities of geothermal exploitation, this research includes a deep Electrical-Resistivity Tomography (ERT) model obtained from the geophysical prospecting on a certain study case, with an already known favorable geological setting and thermal evidence. From the geophysical campaign tests, the geological and thermal characterization of the area was performed to finally evaluate the possible geothermal use. Despite the preliminary discovery of anomalous underground temperatures, there is, in the study area, a lack of knowledge about the distribution of the geothermal resource at depth and the viability of its extraction. Based on this, the importance of this research lies in achieving a greater characterization of the underground area, being the final aim of the work is to provide a new basis for possible future geothermal exploitation. In turn, this research pursues to highlight the benefits of geophysics when characterizing the underground to evaluate the proper geothermal exploitation and optimize the configuration of the global well field and the corresponding elements of the system. In this context, the present paper is organized as follows: firstly, information about the geological and geothermal conditions of the study area and the geophysical technique implemented in the prospecting campaign is included. Then, the results of the geophysical tests are presented, as well as the discussion of the main achievements derived from the experimental phase. As a final section, the paper includes the conclusions and future perspectives of applications.

2. Preliminary Study-Area Characterization

2.1. Geology and Structural Setting

As previously mentioned, the objective of this study is the analysis of the subsoil characteristics in order to clarify the possibilities of future deep–medium geothermal energy exploitation. In this context, the study focuses on a certain area located in the autonomous community of Granada (Spain), in which an electrical-resistivity tomography survey has been performed. The following Figure 1 describes the location of the study area selected in the present research.

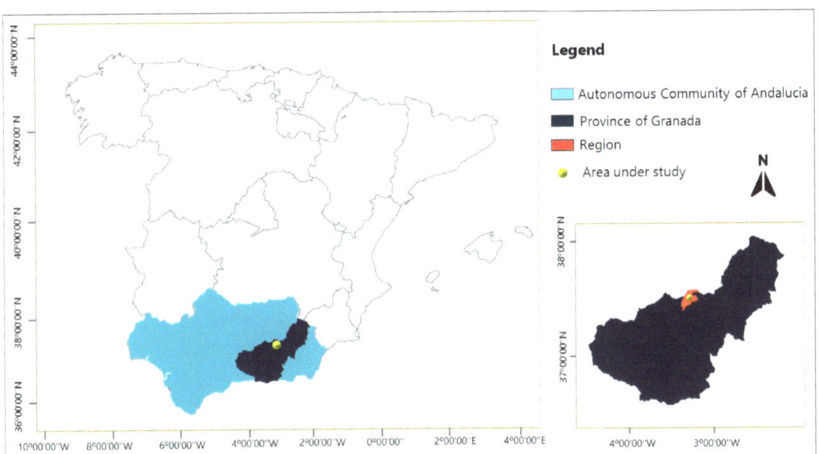

Figure 1. Location of the study area included in the evaluation of this research.

The selection of the location under study is based on the structural and geological characterization that indicated a possible formation of interest from the point of view of energy use. The area under study is located within a series of mountainous alignments to the south of the Guadalquivir Valley, known as the Baetic Mountains and, more specifically, in the region of the "Montes Orientales". The area suffered tectonic phenomena on a continental scale during most of the Mesozoic and Tertiary, related to the opening of the Atlantic and Tethys, as well as the collision of the European and African plates [23,24]. In particular, the area included in this research is located in the sub-Baetic zone, characterized by presenting practically continuous sedimentation between the Triassic and the lower Miocene. Three main domains are established from north to south in this context:

- External sub-Baetic, corresponding to an area in which limestone materials predominate, acquiring great development levels of condensation.
- Medium sub-Baetic, predominating marly materials that appear as characteristic lithologies, radiolarites, and submarine volcanic rocks.
- Internal sub-Baetic that covers a relatively slightly subsiding sector in which exclusively limestone materials outcrop.

Beyond the described sub-Baetic level, Neogene and Quaternary materials are also located in the analysed environs that appear as alluvial deposits. Figure 2 presents the geological levels that constitute the area under study [25].

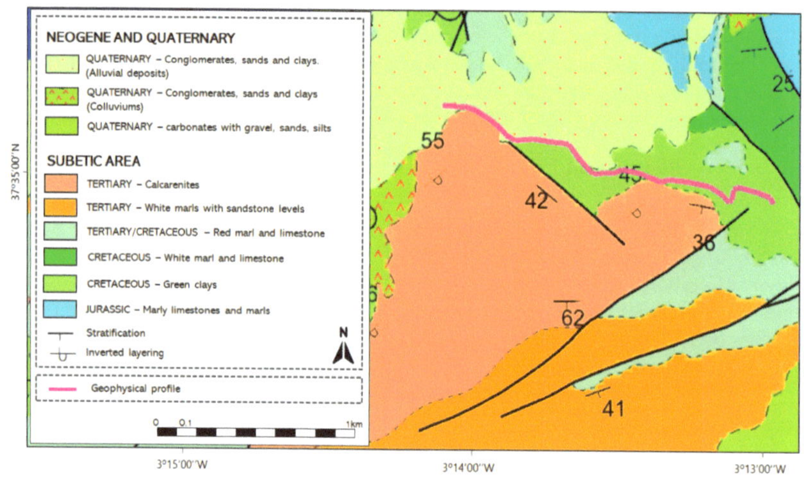

Figure 2. Geological setting of the area in which the geophysical survey was performed.

2.2. Previous Geothermal Evidences

In addition to the geological characterization presented above, in the vicinity of the area where the geophysical campaign has been conducted, there is an existing drilling whose open-access information is also relevant to the aim of this study. The 219 m borehole (included in Figure 3) was drilled by percussion with a downhole trephine and allowed for the deduction of the lithological column described in Table 2 [26].

During the prolonged pumping in the drilling, several measurements of the temperature of the extracted water were conducted. It was stabilized at 35.2 °C, which could preliminarily indicate the existence of a low-temperature geothermal resource (placing this temperature above the average of the place by around 15 °C). The interpretation of the borehole tests and the results of the pumping test also allowed for the conclusion that the storage formation has a calculated transmissivity of 600 m^2/day and an estimated exploitation flow of at least 60 L/s, with maximum values around 80–90 L/s.

Figure 3. Arrangement of the ERT profile in the area under study (at the **top**) execution of works in the field (at the **bottom**).

Table 2. Description of the lithological column of the existing borehole located in the study area.

Length	Geological Description
0–170 m	Alternations of light and white colors marls assigned to the middle–lower Cretaceous
170–198 m	Alternation of sandy and marl sections with limestone From the Lower Cretaceous and Upper Jurassic
198–214 m	Marls and white limestones from the Middle Jurassic
214–219 m	Tabled limestone and karstified gray dolomitic limestone

3. Materials and Methods

3.1. Geophysical Surveys-ERT

3.1.1. ERT for Geothermal Characterization Review

Based on the description of the previous Sections 2.1 and 2.2, this research aims to determine the in-depth characterization of the structure and arrangement of the subsoil materials and analyze the lateral continuity of the formation of interest, with special attention to the potential presence of fractures. With these objectives in mind, the geophysical-prospecting campaign has been raised considering the partial knowledge of the ground in the study area and the need to reach sufficient depth levels for the investigation here pursued. From the commented initial statements and, as previously mentioned, the electrical-resistivity tomography technique, it was considered as the potential method for the underground evaluation of this work.

ERT survey is widely implemented for mapping the location of potential areas for groundwater, minerals, or geothermal use [27,28]. In the specific geothermal field, different studies have focused on the delineation of geothermal reservoirs and/or associated structures such as fractures or faults as a preliminary prerequisite for a successful geothermal exploration [29]. In the evaluation of large and known geothermal areas, ERT profiles have proven to be a suitable method for obtaining a high spatial variability, where the potential zone is clearly distinguishable from those with different resistivity [30–32]. In the case of high- and medium-enthalpy geothermal systems, these present great variations in their resistivity/conductivity structure, usually associated with the occurrence of fluids but also with the presence of anomalous concentrations of hydrothermal minerals. The existence of particular fluids, such as saline fluids or meteoric water in rocks, results in enhanced electrical conductivity to a greater or lesser extent, depending on the characteristics of this fluid [33–35]. In these cases, ERT methods are an important source of information, but they also constitute a great help in those surveys where the initial thermal evidence is not so clear, but a possible state of fracturing or structural disposition could indicate the potentiality of geothermal exploitation at different scales. In the case of low- and very-low-enthalpy geothermal resources, these geophysical tools are also extremely valuable for the characterization and distribution of the ground materials and the subsequent design of the well field [18,19].

3.1.2. Fundamentals of the Method and Application on the Study Case

ETR technique provides a subsurface geoelectrical characterization by the measurement of the apparent resistivity from a tetra-electrode device, injecting current of a known intensity into two electrodes called "A" and "B" and automatically recording the potential difference between the other two electrodes "M" and "N". The process is consequently repeated by automatically varying the distances between the pairs of electrodes so that the apparent resistivity is obtained in multiple positions and levels (n). Data are subsequently processed by means of mathematical inversion algorithms, obtaining an image of resistivities and real depths of the subsoil. The depth of investigation is in fact a function of the ability of the material to identify and measure the electric potential between electrodes [36].

The methodology is based on the contrast of resistivities obtained that allows for the differentiation of the subsoil materials based on their electrical behavior; that is, their apparent resistivity value (after the inversion of the field data). However, the subsurface resistivity of a particular area can be affected by different factors such as the porosity and pore structure of the rock formations, the content of water or steam, salinity, pressure, temperature, or any other alteration between water–rock. In this sense, the pore volume, regarding the total rock volume ratio, the geometric arrangement of the pores (formation factor), the ratio of water-filled pores to empty pores, or the resistivity/conductivity of the fluid that is filling the pores [37], is particularly influential.

In the case of the present research, field works consisted of the execution of an electrical tomography profile of 2000 m in length using a hose with 21 electrodes with a separation of 100 m, and a direct and inverse pole–dipole recording device (Figure 3). Data were collected using the commercial equipment Syscal Pro multi-electrode imaging system with an automatic injection range with a 1200 W AC/DC converter. The array configuration was set with the aim of obtaining a strong signal, high-penetration capability, and high-density data. Since the data quality during the ERT acquisition depends on factors such as the noise or the resistivity of materials, the use of a high-power transmitter and an AC/CD converter increases the effective working voltage and contributes to their mitigation. One more influential factor is contact resistance, which is a function of the resistivity of the most superficial layer of the ground. If contact resistances are not allowed, longer and/or multiple linked electrodes are used at the electrode position, and if required, the galvanic contact of electrodes with the ground can be improved by adding a saltwater solution [38]. It is also convenient to mention that the location of each electrode was defined by using a handheld Global Positioning System (GPS) with a horizontal accuracy of 3 m. For the GPS coordinates, ellipsoidal altitudes were also determined since elevation values are required for the subsequent data inversion.

3.1.3. Inversion-Model Approach

Once the ERT raw data on the field (binary) were acquired, these were converted into ASCII format to numerically process the field datasets. For removing the outlier data points,

X2IPI software (version 5.19) was used following the criteria of unmeasured intensities of those values lower than 100 mA [39]. In the next step, topography information was assigned to each node, and the apparent resistivity data were inverted using a linearized least-squares algorithm in order to obtain the inverted resistivity models [40].

Finally, RES2DINV software (version 5.0) was implemented for ERT data modeling. The tool is based on a finite element modeling or a difference algorithm capable of providing the forward modeling of the voltage response to the current injection. The models of resistivity developed by the software are then divided into different rectangular blocks with specific resistivity values coming from the field-measuring process. Data of apparent resistivity are finally presented as a pseudo-section, in which the apparent resistivity values are assigned to a predefined location in function of the type of array [41].

The selected inversion procedure was the smoothness-constrained least-squares optimization method, or the smooth L2 norm (function of a damping factor), which allows for the minimization of the sum of squares between the calculated apparent resistivity values and those observed, producing smooth fluctuations in resistivity within the inversion model. The process of inversion begins with the initial model parameters and the damping factors that are refined through an iterative process.

4. Results

4.1. 2D ERT Profile

Figure 4 shows the results obtained in the 2D ERT profile performed in the area under study. The geoelectrical model also includes the synthetic column crossed by the existing drilling together with the distinction of three main geological horizons and the location of the fault's structures in line with the geological characterization established in Section 2.

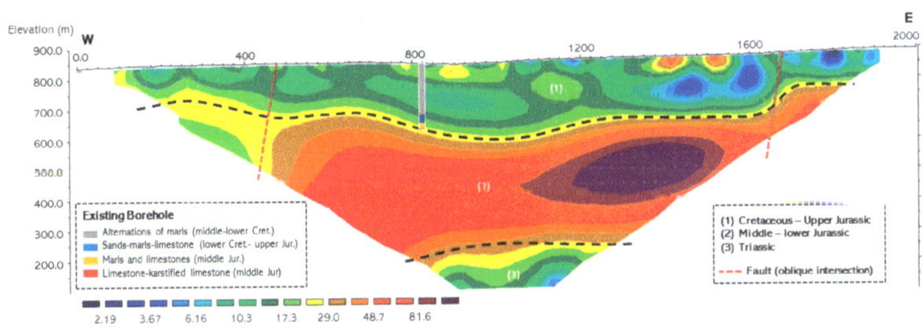

Figure 4. Electrical-resistivity tomography 2D profile in the study area. Length: 2000 m, Interelectrode span: 100 m, Number of electrodes: 21, Dispositive registration: Pole–Dipole, RMS error: 10.3%.

In the 2D model of the above Figure 4, three layers can be distinguished with the following characteristics:

- A first outcropping surface layer (1) of a relatively conductive nature, corresponding to Cretaceous and Upper Jurassic materials mainly constituted by alternations of sandy, limestone, and marly materials. As observed in the 2D profile, Layer (1) presents a more conductive character in the extreme east, indicating the majority presence of clays and marls. Regarding the thickness of the layer, it ranges between 120–250 m.
- An intermediate resistive layer (2), constituted by carbonate materials from the middle and lower Jurassic. This layer has a significant thickness higher than 300 m and presents two main fracture areas (also included in Figure 4):
    - ○ A fracture located around Meter 450 of the 2D profile. The area is defined by the lateral change in resistivity observed in the model that descends at that point in a westerly direction. It should be noted that this fracture context agrees with the fault mapped by the consulted geological database [23], which intersects the trace of the profile at this point, but with a very oblique character.

This factor could influence the resistivity values obtained from the geophysical prospecting campaign.

   ○ A fracture formation appearing approximately at Meter 1700 of the ERT profile, defined by the jump and elevation observed at the top of layer (2), estimated at about 80 m.

- A final conductive basal layer (3), determined by the clear decrease in resistivity, which could indicate the presence of Triassic materials.

In general, the described results indicate that there is an appropriate correlation between the data derived from the existing borehole and the geoelectrical model obtained from geophysics.

4.2. Pseudo-3D ERT Model

In addition to the previous 2D profile, a three-dimensional model has been obtained using Oasis Montaj software (Seequent). Once the profile into the tool is exported, it performs an interpolation of the areas without data from the tomography profile, located in the basal part of the ends. The final product is a 3D geoelectrical model that enables an in-depth analysis of the whole ground distribution [42]. These 3D models are included in the following Figures 5–7 in which it is possible to observe the general structure of the study area in relation to its topography and geological cartography.

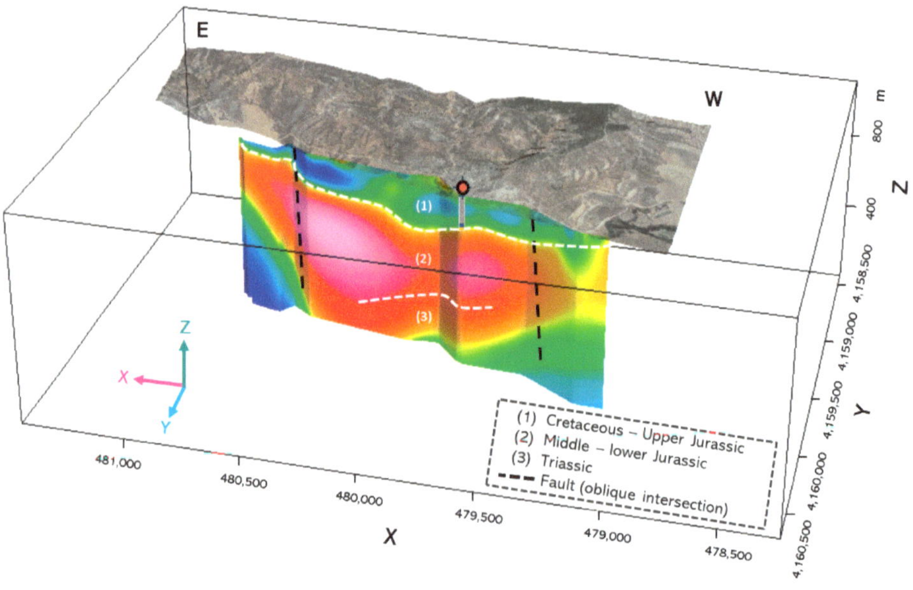

Figure 5. Pseudo 3D model with the digital Terrain Model (DTM) and the orthophoto. View from the north.

The previous Figures 5–7 show the tomography profile obtained from the geophysical campaign (depth information) overlaid with three surface visualization modes: according to the digital model of the terrain and the orthophoto, the single digital terrain model and the digital terrain model together with the geology and family of faults in the area. In all of them, the location of the existing borehole has also been included in order to facilitate the interpretation of the results.

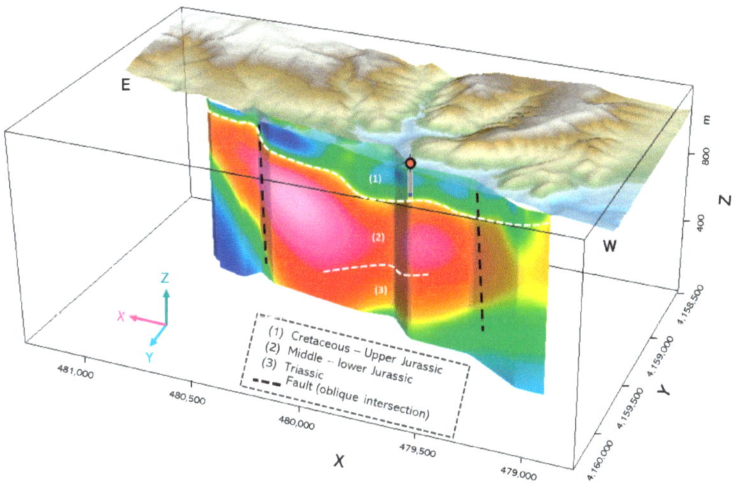

Figure 6. Pseudo-3D model with the Digital Terrain Model (DTM). View from the north.

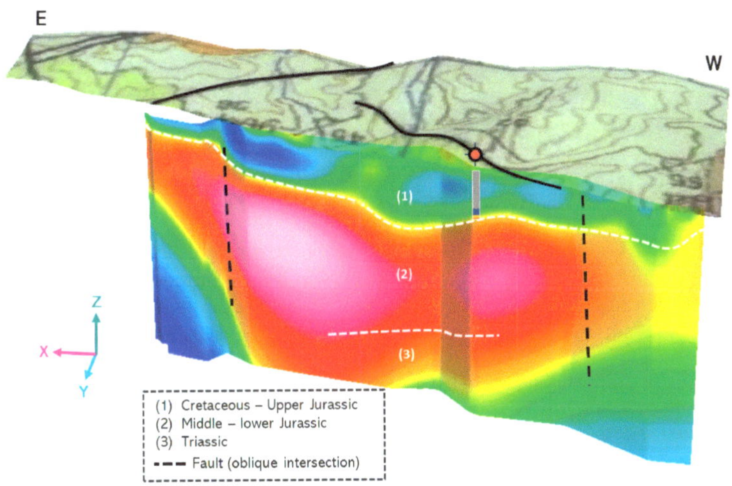

Figure 7. Pseudo-3D model with DTM and geology, highlighting the main mapped faults. View from the north.

5. Discussion

5.1. Model Validation

In certain exploration areas, there is great variability between the geological conditions and the underground distribution. In this sense, the application of geophysics is fundamental for understanding the existing structures and formations in depth. The present research investigates the potential of ERT for the exploration of low–medium enthalpy geothermal resources.

The implementation of an ERT 2000 m long profile has allowed for the penetration of the subsoil up to levels of around 800 m, meaning it is an important source of information about the composition of the materials that make up the subsoil and its distribution in the horizons. The obtained geoelectrical model must represent the subsurface and, at the same time, provide a good fit to the surface geology and lithostratigraphic units based on the existing geological cartography and the known fracture formations and faults.

According to the 1:50,000 scale cartographic layer produced by the IGME (Instituto Geologico y Minero de España) in the area under study, two families of fractures intersect the profile at the eastern and western ends of it. This source of information provides an initial basis for the nature of the characteristic geological formations of the site, which, together with the lithological column known from the borehole existing in the area, represents an important approximation to the characterization of the subsoil. The interpretation of the ERT results agrees with the existing information in the first levels of the ground, also providing additional documentation of how the geological structures are distributed in depth. Through the geophysical campaign, it has been possible to corroborate the arrangement of the fault formations at deeper levels and to characterize the subsoil at greater levels of depth. The results of the survey denote the existence of a potential formation constituted by carbonate materials from the middle and lower Jurassic with a thickness greater than 300 m, which is also crossed by the families of fractures. The set of information provided by the tests and verified by the known structures at more superficial levels makes it possible to accurately plan the most optimal location for possible geothermal exploitation in the analyzed area. The data known from the drilling report of the borehole in the area also confirm (although in much less detail) a significant coincidence in the division of layers in depth.

5.2. Geothermal Exploitation

As commented before, based on the results obtained in the geophysical model, it is possible to establish a preliminary approximation of the most appropriate scenario for future geothermal exploitation. The approach followed here is to evaluate a possible opportunity to semi-directly capture the geothermal flow that is suspected to exist in the underground study area. Due to the limitations of the geophysical techniques (which can only be solved with direct drilling), there is not total certainty about the behavior of the phenomenon, but this section tries to provide a possible exploitation pattern according to the known in-depth information.

Following the geological distribution in the ground, the geothermal drilling should reach the formation (2) of Figures 4–7 in which carbonate materials are present. These layers, characterized in their most superficial part by a certain degree of karstification, constitute a proper environment for the extraction of the required water flow by means of geothermal use. Specifically, the projected borehole should be located at the extreme west of the profile (Figure 8) in the vicinity of the fault structure, and where the formation of interest (2) does not present resistivity values as high as in the case of the existing borehole.

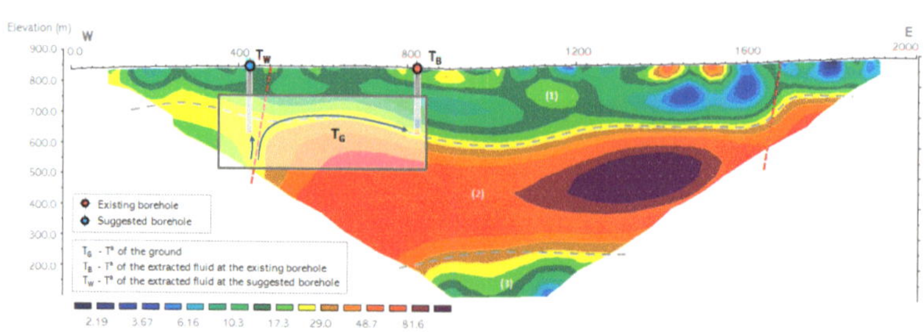

Figure 8. Location of the proposed drilling for possible geothermal exploitation.

Once the potential location for a possible geothermal use is established, an estimation of the temperature of the fluid in the considered area and the possible thermal extraction is presented in this subsection. For this, the probable flow path from the potential area (in the fault) to the existing drilling will be evaluated. Taking into account the geological distribution analyzed in this study, the probable circulation layer should be the one marked in yellow (Figure 8) that communicates the fracture with the lower end of the existing borehole. In this way, and based on the scale of Figure 8, the circulation area is approximately 400 m long and 40 m thick. From this information, and considering the porosity of the limestone formation of the reservoir, the real diameter through which the flow circulates up to the borehole is obtained (D_F of Table 3).

As a second step, it is necessary to define the initial thermal conditions of the site, such as the average temperature of the ground at the depth considered (220 m as the existing drilling). Based on previous studies, the area is not characterized by known significant geothermal anomalies, having a normal geothermal gradient of around 3 °C/100 m and with temperatures within the average values in the original geological environment [43]. The estimated ground temperature at the considered level can be also found in Table 3 as T_G.

Table 3. Principal site and fluid parameters required for the thermal characterization of the site [44].

Main Site Parameters	
Flow diameter (D_F)	3.2 m
Ground Temperature (T_G)	19.5 °C
Temperature of the water borehole (T_B)	35.2 °C
Length of the area (L_A)	400 m
Thickness of the area (T_A)	40 m
Considered depth (D_P)	220 m
Fluid Properties	
Kinematic viscosity (υ)	1×10^{-6} m^2/s
Dynamic viscosity (μ)	1×10^{-3} kg/m·s
Specific heat (c_p)	4184 J/kg·K
Thermal conductivity (k)	0.6 W/m·K

Based on the above information, the heat flux by forced convection can be evaluated, for which different parameters must be defined. The first one is the dimensionless Reynolds number (Re), required to categorize the fluid of the system and to confirm the laminar behavior of the flow (Equation (1)) [45].

$$Re = \frac{V \cdot D_F}{\upsilon} \tag{1}$$

where V is the average flow velocity (m/s).

In the present study case, the flow velocity (V) is estimated according to measurements obtained from the existing borehole, such as the transmissivity of 600 m^2/day already mentioned in Section 2.2 and the analyzed layer thickness (T_A), obtaining a maximum flow velocity of around 15 m/day. Regarding the kinematic viscosity (υ), it is consulted in standard databases for water at the temperature of the medium considered (Table 3) [43]. With all these values, a Reynolds number of 555.56 is obtained, denoting the laminar nature of the fluid in the analysed conditions of the environment.

The following parameters to calculate are the Prandtl number (Pr), which expresses a dimensionless quantity for assessing the relation between momentum transport and thermal transport capacity of the fluid, and the Nusselt number (Nu) that is basically a function of Re and Pr numbers. Both numbers are defined in Equations (2) and (3) [45].

$$Pr = \frac{\mu \cdot C_p}{k} \tag{2}$$

$$Nu = 3.66 + \frac{0.065 \cdot \left(\frac{D_F}{L_A}\right) \cdot Re \cdot Pr}{1 + 0.04 \cdot \left[\left(\frac{D_F}{L_A}\right) \cdot Re \cdot Pr\right]^{2/3}} \quad (3)$$

Considering again the standard values of μ, c_p, and k for the fluid at the analyzed conditions (Table 3) [46], Pr and the Nu numbers are estimated as 6.97 and 5.26, respectively. It is convenient to mention that Equation (3) provides an approximation to the calculation of Nu, considering that the value obtained is greater than 3.66 as the one here presented.

Once defined, the previous parameters, the thermal convection coefficient (h) (W/m^2·K) for forced convection, is expressed as follows (Equation (4)).

$$h = \frac{k}{D_F} \cdot Nu \quad (4)$$

Substituting the corresponding values, the convection coefficient h for the study conditions is 0.99 W/m^2·K. Convective heat transfer occurs from the moving of the fluid through the considered formation, according to what is known as Newton's Law of Cooling. In this way, from the calculated convection coefficient, the temperature of the water (T_W) in the suggested location of the fracture can be calculated by applying Equation (5) [47].

$$T_W = T_{G-} \left[\frac{T_G - T_B}{e^{-\frac{h \cdot A_S}{\dot{m} \cdot c_p}}} \right] \quad (5)$$

where A_s (m^2) is the area through which the fluid circulates in the formation and $\dot{m}$ (kg/s) is the mass flow rate.

Considering the geometry of the formation, as well as the density and the estimated velocity of the fluid, A_s and $\dot{m}$ can be directly obtained, being the values 3516.80 m^2 and 1.40 kg/s, respectively. Finally, applying the previous Equation (5), the water temperature in the area suggested in this study could reach the value of 48.89 °C.

On top of the above, the specific energy (E) that the increase in temperature supposes in both conditions (in the existing drilling and in the suggested one) is calculated applying the following equation.

$$E = c_p \cdot \Delta T \cdot m \quad (6)$$

where m (kg/h) is the mass of the circulating fluid, which can be calculated from the estimated flow in the existing drilling (at least 60 L/s) that, considering the density of the fluid, is means 216,000 kg/h. ΔT (°C) is the temperature increase between the ground and the temperature of the water in each assumption. These values, and the final result of applying Equation (6), are included in Table 4.

Table 4. Increase in temperature and specific energy achieved in each of the considered scenarios.

Scenario	ΔT (°C)	E (J)	E (th)
Existing drilling	15.70	1.42×10^{10}	3388.93
Suggested drilling	29.39	2.66×10^{10}	6343.99

From the previous values of Table 4, it is easily observable how the extraction of water in the area suggested in the present research supposes an increase in temperature of almost double that currently achieved in the existing borehole of the area. All this also means doubling the specific energy of the site and denoting the presence of a relevant geothermal resource of significant importance for its future use.

In relation to the previous calculations of the upwelling temperature, it should be clarified that this is one of the possible scenarios regarding the origin of the thermal anomaly measured in the well. For a deeper characterization of the resource, it would be necessary to take into account possible deviations in the geometry and properties of the geological structure that constitute the aquifer (error of the inversion process of geophysical data), or possible thermal contributions not considered in the direction of the flow prior to the emerging fracture.

6. Conclusions

In different regions of Spain, the lack of precise information on the resource and its possibilities of exploitation constitute a barrier preventing geothermal energy from making a greater contribution to meeting energy demand at present. The present study applies deep electrical-resistivity tomography to characterize a possible geothermal site. The information about the subsoil obtained from the resistivity data is in agreement with the geological structures known from the existing borehole located in the vicinity of the studied area. Beyond this contribution, the geophysical campaign has allowed to know the distribution of the geological formations and structures at greater depth and define the ideal location for better geothermal resource exploitation. In this way, and by locating the possible energy source in depth through the structuring in the form of a fault, the extraction of water in the proposed area allows us to achieve an estimated temperature of 48 °C with an energy use of practically double the current extraction in the existing borehole.

Based on all that has been evaluated in this investigation, future geothermal exploitation initiatives in the study area will be precisely planned, minimizing the possibilities of error and facilitating the corresponding exploration and implementation tasks linked to the geothermal-extraction system. Regarding future research, and since it is a promising area, it would be interesting to address the analysis of the formations that give rise to thermal anomalies on the surface through the use of more extensive geophysical techniques. In this sense, the application of the magnetotelluric method could be advisable to image the subsurface electrical resistivity and provide (by the use of the Earth's naturally occurring electromagnetic fields) useful information about the lateral and vertical resistivity variation. The investigation depth of this technique can reach several tens of kilometers in function on rock resistivity, making it a technology of great value to be applied in areas such as the one evaluated in this research, making it also possible to determine with greater precision the scope of the resource and its possibilities of use as a medium or a high-enthalpy geothermal system.

Author Contributions: Conceptualization, C.S.B. and I.M.N.; methodology, C.S.B.; software, P.C. and J.C.; validation, A.F.M. and D.G.-A.; formal analysis, C.S.B., I.M.N. and M.Á.M.-G.; investigation, C.S.B.; resources, P.C., D.P. and J.C.; data curation, P.C., D.P. and J.C.; writing—original draft preparation, C.S.B.; writing—review and editing, C.S.B.; visualization, M.Á.M.-G.; supervision, A.F.M. and D.G.-A.; project administration, C.S.B.; funding acquisition, A.F.M., C.S.B. and M.Á.M.-G. All authors have read and agreed to the published version of the manuscript.

Funding: This research was supported by the research projects support program and cofinanced by the European Regional Development Fund (ERDF) under the order of the Ministry of Education from Castilla y León (project ref.: SA102P20). "Development and integration of a new district heating system using low-energy geothermal energy (GEO-DISTRICT 3.0)" and the European Union's Horizon 2020 research and innovation programme under grant agreement nº 101091885 (Mine.io project). M.Á.M.-G. and C.S.B. acknowledge the grant RYC2021-034813-I and RYC2021-034720-I, respectively, funded by MCIN/AEI/10.13039/501100011033 and by European Union "NextGenerationEU"/PRTR.

Data Availability Statement: Data will be available when required.

Acknowledgments: Authors would also like to thank the Department of Cartographic and Land Engineering of the Higher Polytechnic School of Avila, University of Salamanca for allowing us to use their facilities and their collaboration during the experimental phase of this research.

Conflicts of Interest: The authors declare no conflicts of interest.

Nomenclature

Acronyms	
GHP	Geothermal Heat Pump
GSHP	Ground-Source Heat Pump
GWHP	Ground-Water Heat Pump
NZEB	Nearly Zero-Energy Building
ERT	Electrical-Resistivity Tomography
GPS	Global Positioning System
IGME	Instituto Geológico y Minero de España
D_F	Flow diameter
T_G	Ground Temperature
T_B	Temperature of the water borehole
T_W	Temperature of the water
L_A	Length of the area
T_A	Thickness of the area
D_P	Considered depth
υ	Kinematic viscosity
v	Flow velocity
μ	Dynamic viscosity
c_p	Specific heat
k	Thermal conductivity
Re	Reynolds number
Pr	Prandtl number
Nu	Nusselt number
h	Thermal convection coefficient
A_s	Area of circulation of the fluid in the formation
$\dot{m}$	Mass flow rate
E	Specific energy
m	Mass of the circulating fluid

References

1. Fridleifsson, I.B. Geothermal energy for the benefit of the people. *Renew. Sustain. Energy Rev.* **2001**, *5*, 299–312. [CrossRef]
2. García-Gil, A.; Goetzl, G.; Kłonowski, M.R.; Borovic, S.; Boon, D.P.; Abesser, C.; Janza, M.; Herms, I.; Petitclerc, E.; Erlström, M.; et al. Governance of shallow geothermal energy resources. *Energy Policy* **2020**, *138*, 111283. [CrossRef]
3. Dickson, M.H.; Fanelli, M. *Geothermal Energy: Utilization and Technology*; Routledge: London, UK, 2013; ISBN 978-1-84407-184-5.
4. Lund, J.W.; Toth, A.N. Direct utilization of geothermal energy 2020 worldwide review. *Geothermics* **2021**, *90*, 101915. [CrossRef]
5. Sáez Blázquez, C.; Martín Nieto, I.; Farfán Martín, A.; González-Aguilera, D.; Carrasco García, P. Comparative analysis of different methodologies used to estimate the ground thermal conductivity in low enthalpy geothermal systems. *Energies* **2019**, *12*, 1672. [CrossRef]
6. Limberger, J.; Boxem, T.; Pluymaekers, M.; Bruhn, D.; Manzella, A.; Calcagno, P.; Beekman, F.; Cloetingh, S.; van Wees, J.-D. Geothermal energy in deep aquifers: A global assessment of the resource base for direct heat utilization. *Renew. Sustain. Energy Rev.* **2018**, *82*, 961–975. [CrossRef]
7. Rybach, L. Global Status, Development and Prospects of Shallow and Deep Geothermal Energy. *Int. J. Terr. Heat Flow Appl. Geotherm.* **2022**, *5*, 20–25. [CrossRef]
8. Lund, J.W.; Huttrer, G.W.; Toth, A.N. Characteristics and trends in geothermal development and use, 1995 to 2020. *Geothermics* **2022**, *105*, 102522. [CrossRef]
9. Blázquez, C.S.; Borge-Diez, D.; Nieto, I.M.; Martín, A.F.; González-Aguilera, D. Multi-parametric evaluation of electrical, biogas and natural gas geothermal source heat pumps. *Renew. Energy* **2021**, *163*, 1682–1691. [CrossRef]
10. D'Agostino, D.; Mazzarella, L. What is a Nearly zero energy building? Overview, implementation and comparison of definitions. *J. Build. Eng.* **2019**, *21*, 200–212. [CrossRef]
11. Blázquez, C.S.; Borge-Diez, D.; Nieto, I.M.; Martín, A.F.; González-Aguilera, D. Technical optimization of the energy supply in geothermal heat pumps. *Geothermics* **2019**, *81*, 133–142. [CrossRef]
12. Magrini, A.; Lentini, G.; Cuman, S.; Bodrato, A.; Marenco, L. From nearly zero energy buildings (NZEB) to positive energy buildings (PEB): The next challenge-The most recent European trends with some notes on the energy analysis of a forerunner PEB example. *Dev. Built Environ.* **2020**, *3*, 100019. [CrossRef]
13. Sorman, A.H.; García-Muros, X.; Pizarro-Irizar, C.; González-Eguino, M. Lost (and found) in Transition: Expert stakeholder insights on low-carbon energy transitions in Spain. *Energy Res. Soc. Sci.* **2020**, *64*, 101414. [CrossRef]

14. Arrizabalaga, I.; De Gregorio, M.; De Santiago, C.; García de la Noceda, C.; Pérez, P.; Urchueguía, J.F. Country update for the Spanish geothermal sector. In Proceedings of the World Geothermal Congress, Melbourne, Australia, 19–24 April 2015; pp. 19–24.
15. Arrizabalaga, I.; De Gregorio, M.; De Santiago, C.; García de la Noceda, C.; Pérez, P.; Urchueguía, J.F. Country update for the Spanish geothermal sector. In Proceedings of the World Geothermal Congress, Reykjavik, Iceland, 27 April–1 May 2020; p. 12.
16. Vivas, C.; Salehi, S.; Tuttle, J.D.; Rickard, B. Challenges and opportunities of geothermal drilling for renewable energy generation. *GRC Trans.* **2020**, *44*, 904–918.
17. Anderson, A.; Rezaie, B. Geothermal technology: Trends and potential role in a sustainable future. *Appl. Energy* **2019**, *248*, 18–34. [CrossRef]
18. Blázquez, C.S.; Nieto, I.M.; González, M.Á.M.; García, P.C.; Martín, A.F.; González-Aguilera, D. Geophysical exploration for shallow geothermal applications: A case study in Artà, (Balearic Islands, Spain). *Geothermics* **2022**, *105*, 102517. [CrossRef]
19. Sáez Blázquez, C.; Carrasco García, P.; Nieto, I.M.; Maté-González, M.Á.; Martín, A.F.; González-Aguilera, D. Characterizing geological heterogeneities for geothermal purposes through combined geophysical prospecting methods. *Remote Sens.* **2020**, *12*, 1948. [CrossRef]
20. Carrier, A.; Fischanger, F.; Gance, J.; Cocchiararo, G.; Morelli, G.; Lupi, M. Deep electrical resistivity tomography for the prospection of low-to medium-enthalpy geothermal resources. *Geophys. J. Int.* **2019**, *219*, 2056–2072. [CrossRef]
21. Nieto, I.M.; Carrasco García, P.; Sáez Blázquez, C.; Farfán Martín, A.; González-Aguilera, D.; Carrasco García, J. Geophysical prospecting for geothermal resources in the south of the Duero Basin (Spain). *Energies* **2020**, *13*, 5397. [CrossRef]
22. Ismail, N.; Syukri, M.; Idroes, R. Deep and shallow structures of geothermal Seulawah Agam based on electromagnetic and magnetic data. *Geomate J.* **2019**, *16*, 141–147. [CrossRef]
23. Mohamed, M.A.; Pajares, A.J.M.; Moreno, J.L.; Herrador, M.B.; Álvarez, F.P.; Reyes, J.M.G. Phylogenetic relationships of "Erysimum"(Brassicaceae) from the Baetic Mountains (SE Iberian Peninsula). In *Anales del Jardín Botánico de Madrid*; Real Jardín Botánico: Madrid, Spain, 2014; Volume 71, p. 5. [CrossRef]
24. Sala, M. Baetic Cordillera and Guadalquivir Basin. In *Geomorphology of Europe*; Palgrave: London, UK, 1984; pp. 323–340. [CrossRef]
25. Albir, A.E.; López Olmedo, F.L.; Díaz de Neira, J.A. Mapa y memoria explicativa de la Hoja 970, El Rubio. MAGNA. IGME. 1981.
26. Guzmán, S.J. *Sondeo de Agua Minero Industrial de Origen Termal "Alamedilla 1" Capacity Report*; Junta de Andalucía: Andalusia, Spain, 2015.
27. Lévy, L.; Maurya, P.K.; Byrdina, S.; Vandemeulebrouck, J.; Sigmundsson, F.; Árnason, K.; Ricci, T.; Deldicque, D.; Roger, M.; Gibert, B.; et al. Electrical resistivity tomography and time-domain induced polarization field investigations of geothermal areas at Krafla, Iceland: Comparison to borehole and laboratory frequency-domain electrical observations. *Geophys. J. Int.* **2019**, *218*, 1469–1489. [CrossRef]
28. Singh, S.; Gautam, P.K.; Bagchi, D.; Singh, S.; Kumar, S.; Kannaujiya, S. 2D electrical resistivity imaging for geothermal groundwater characterization and rejuvenation of the Gaurikund hot spring in the Main Central Thrust (MCT) zone of the Garhwal Himalaya, Uttrakhand, India. *Groundw. Sustain. Dev.* **2021**, *15*, 100686. [CrossRef]
29. Kumar, D.; Thiagarajan, S.; Rai, S.N. Deciphering geothermal resources in Deccan Trap region using electrical resistivity tomography technique. *J. Geol. Soc. India* **2011**, *78*, 541–548. [CrossRef]
30. Vanhooren, L.; Fontaine, O.; Caudron, C.; Vrancken, E.; Dekoninck, W.; De Lathauwer, H.; Hermans, T. *Characterization and Monitoring of the Gunnuhver Geothermal Site Using Electrical Methods (No. EGU23-12873)*; Copernicus Meetings; EGU: Vienna, Austria, 2023. [CrossRef]
31. Cheng, Y.; Wang, C.; Da, W.; Kong, Y.; Hu, X. Anatomy of the convective geothermal system from geophysical and hydrochemical data: A case study from the Changshou geothermal field, South China. *Geophysics* **2023**, *88*, 1–43. [CrossRef]
32. Ussher, G.; Harvey, C.; Johnstone, R.; Anderson, E. Understanding the resistivities observed in geothermal systems. In Proceedings of the World Geothermal Congress, Kyushu, Japan, 28 May–10 June 2000; pp. 1915–1920.
33. Marwan, M.; Isa, M.; Idroes, R.; Nursyafira, N.; Idris, S.; Yanis, M.; Ghani, A.A.; Paembonan, A.Y. Geoelectrical Model of Geothermal Spring in ie Jue Seulawah Deriving from 2D Vlf-Em and dc Resistivity Methods. *J. Appl. Eng. Sci.* **2023**, *21*, 59–69. [CrossRef]
34. Abdullah, A.I.; Musa, M.D.T.; Uno, I.; Jayadi, H. Resistivity model of hallow subsurface to find the path of geothermal manifestation in Bora Village of Sigi Regency of Central Sulawesi. *J. Phys. Conf. Ser.* **2019**, *1242*, 012048. [CrossRef]
35. Cumming, W.; Mackie, R. Resistivity imaging of geothermal resources using 1D, 2D and 3D MT inversion and TDEM static shift correction illustrated by a Glass Mountain case history. In Proceedings of the World Geothermal Congress, Bali, Indonasia, 25–30 April 2010; pp. 25–29.
36. Arjwech, R.; Everett, M.E. Application of 2D electrical resistivity tomography to engineering projects: Three case studies. *Songklanakarin J. Sci. Technol.* **2015**, *37*, 675–681.
37. Zhou, B.; Bouzidi, Y.; Ullah, S.; Asim, M. A full-range gradient survey for 2D electrical resistivity tomography. *Near Surf. Geophys.* **2020**, *18*, 609–626. [CrossRef]
38. Carrasco García, P. Avance en Técnicas Geofísicas Para la Caracterización del Subsuelo Mediante Innovación y el Uso de Herramientas de Gestión de Información Espacial. Ph.D. Thesis, Escuela Politécnica Superior de Ávila, Univ de Salamanca, Ávila, Spain, 2013; p. 523.
39. Robain, H.; Bobachev, A. X2IPI Toolbox for 2D DC measurements with SYSCAL equipment. *User manual*, 2002; 25p.

40. Loke, M.H.; Dahlin, T. A comparison of the Gauss–Newton and quasi-Newton methods in resistivity imaging inversion. *J. Appl. Geophys.* **2002**, *49*, 149–162. [CrossRef]
41. Telford, W.M.; Telford, W.M.; Geldart, L.P.; Sheriff, R.E. *Applied Geophysics*; Cambridge University Press: Cambridge, UK, 1990; ISBN 1139642928/9781139642927.
42. Arifin, M.H.; Kayode, J.S.; Ismail, K.I.; Abdullah, M.; Embrandiri, A.; Nazer, S.M.; Azmi, A. Data for the industrial and municipal environmental wastes hazard contaminants assessment with integration of RES2D techniques and Oasis Montaj software. *Data Brief* **2020**, *33*, 106595. [CrossRef]
43. Galindo-Zaldívar, J.; Jabaloy, A.; Serrano, I.; Morales, J.; González-Lodeiro, F.; Torcal, F. Recent and present-day stresses in the Granada Basin (Betic Cordilleras): Example of a late Miocene-present-day extensional basin in a convergent plate boundary. *Tectonics* **1999**, *18*, 686–702. [CrossRef]
44. Mao, Y.; Zhang, Y. Thermal conductivity, shear viscosity and specific heat of rigid water models. *Chem. Phys. Lett.* **2012**, *542*, 37–41. [CrossRef]
45. Abraham, J.P.; Sparrow, E.M.; Tong, J.C.K. Heat transfer in all pipe flow regimes: Laminar, transitional/intermittent, and turbulent. *Int. J. Heat Mass Transf.* **2009**, *52*, 557–563. [CrossRef]
46. Crittenden, J.C.; Trussell, R.R.; Hand, D.W.; Howe, K.J.; Tchobanoglous, G. *MWH's Water Treatment: Principles and Design*, 3rd ed.; John Wiley & Sons: Hoboken, NJ, USA, 1861; ISBN 978-0-470-40593-0.
47. Domenico, P.A.; Palciauskas, V.V. Theoretical analysis of forced convective heat transfer in regional ground-water flow. *Geol. Soc. Am. Bull.* **1973**, *84*, 3803–3814. [CrossRef]

energies

MDPI

Article

Numerical Simulation of Geothermal Reservoir Reconstruction and Heat Extraction System Productivity Evaluation

Jinshou Zhu [1], Zhenpeng Cui [2], Bo Feng [2], Hao Ren [2] and Xin Liu [3,*]

[1] Qinghai Bureau of Geological Survey, Xining 810001, China
[2] College of Environment and Resources, Jilin University, Changchun 130021, China
[3] College of Geosciences and Info-Physics, Central South University, Changsha 410083, China
* Correspondence: 225001034@csu.edu.cn

Abstract: The key to ensuring the economic feasibility of EGS mainly includes two points. On the one hand, it is necessary to ensure the connectivity of the artificial fracture network; on the other hand, it is necessary to determine the most efficient geothermal energy exploitation mode. Most previous studies have only focused on one of the points. To restitute the entire geothermal energy development process, the two parts should be combined to conduct research. In this study, a random fractured medium model was established based on the TOUGH2-BIOT simulation program and the whole process of reservoir stimulation was analyzed. According to the results of reservoir stimulation, different geothermal energy exploitation schemes are set up, and the heat transfer efficiency of the conventional double vertical wells, the horizontal wells, and the double-pipe heat exchange system are comparatively analyzed. The results show that reservoir reconstruction is mainly divided into three stages: In the first stage, the hydraulic aperture of the conducting fractures reaches the maximum value; in the second stage, the non-conductive fractures overcome the in situ stress and become conducting fractures; in the third stage, the rock in the reservoir undergoes shear failure, the fractures expand and connect, and finally, a fracture network is formed. After each stage, the volume of the enhanced permeability area is approximately 10,000, 21,000, and 33,000 m^3, respectively. After 30 years of exploitation, the outlet temperature and thermal power output of conventional double vertical wells are the highest, while the horizontal wells have the highest heat extraction ratio. The temperature of a production well in the conventional double vertical wells model, horizontal wells, and double-pipe heat exchange system is 101 °C, 93.4 °C, and 91.6 °C, a decrease of 41.2%, 45.7%, and 46.7%, respectively. The thermal power output is 6.67 MW, 6.31 MW, and 6.1 MW, a decrease of 39.4%, 42.6%, and 44.5%, respectively. The heat extraction ratio of the horizontal wells is 2% higher than the double-pipe heat exchange system and 6.5% higher than the conventional double vertical wells.

Keywords: enhanced geothermal system; reservoir reconstruction; productivity evaluation; TOUGH2-BIOT; numerical simulation

1. Introduction

The energy structure dominated by traditional fossil energy has placed huge pressure on the environment [1]. Countries are always looking for environmentally friendly, energy-efficient, and, especially, renewable energy sources to reduce the proportion of fossil fuels in the energy structure [2]. Geothermal energy has received increasing attention with regard to its green and renewable resource with abundant reserve, and its commercial exploitation is believed to be an effective option to achieve the targets of "carbon peaks" and "carbon neutral" [3,4]. Therefore, the effective development of deep geothermal resources has become a research focus in the geothermal industry.

Hot dry rock (HDR) represents the largest portion of the geothermal energy reserve. According to statistics, the total HDR geothermal resources buried at 3–10 km is 2.5 × 10^{25} J (equivalent to 860 trillion tons of standard coal), with the energy contained equivalent to 30 times the energy contained in all the oil, gas, and coal in the world combined [5,6]. However, due to HDR being dense and hard, and the permeability extremely low, the degree of exploitation is the lowest among geothermal resources. Therefore, in the development and utilization of dry heat geothermal resources, hydraulic fracturing is required to form artificial fracture networks, with heat extracted by injecting flow in water injection wells, circulated in the fracture network, and collected in production wells to form an enhanced geothermal system (EGS) [7,8]. However, the development of HDR geothermal resources exploitation is an extremely complex process, generally expressed as a thermal-hydrological-mechanical (THM) coupling problem. Considering the difficulty of direct monitoring and research on deep underground space, a numerical simulation is an efficient way of solving multi-field coupling problems [9].

The key to ensuring the economic feasibility of EGS mainly includes two points. On the one hand, it is necessary to ensure the connectivity of the artificial fracture network. In the process of reservoir stimulation, the uncontrollability of fracture development is the main reason affecting the connectivity of artificial fracture networks [10]. The fracture system in EGS could be represented by two methods, namely, the discrete fracture network model and the equivalent continuous porous media model [11]. Based on these methods, many numerical models have been established to study the reservoir stimulation of EGS. By establishing a fractured porous media model, Rutqvistz et al. analyzed the effect of stress changes on fracture development in the process of reservoir stimulation [12]. Lei et al. analyzed the changes in pressure, temperature, and stress inside EGS in the stage of fracture development by establishing an equivalent porous media model [13]. Lu et al. and Ghassemi et al. proposed a fully coupled equivalent porous medium model of THM based on rock mass deformation, shear expansion, and fracture expansion when fluid seepage occurs inside fractures, which can quantitatively predict the complex fracture reconstruction process in the fracture system [14,15]. Saeed et al. developed stochastic discrete fracture networks (DFN) to mimic the reservoir permeability behavior based on experimental and field data [16,17].

On the other hand, after establishing the artificial fracture network, the exploitation efficiency of geothermal energy will be affected by different wellbore layouts. The heat transfer between the geothermal reservoir and fluid medium in the EGS is described by two models: local thermal equilibrium and local thermal non-equilibrium. Research shows that compared with other wellbore layout methods, the traditional double vertical wells have the highest outlet temperature, the horizontal wells have the highest heat power output and heat extraction rate, and the double-pipe heat exchange system has the lowest reservoir reconstruction cost and better fracture network connectivity [18–22].

Although the previous numerical simulation studies on EGS performance are extensive, fractures in the geothermal reservoir are randomly distributed, the dip angle and dip tendency of different fractures are different, and the directionality of the fractures needs to be considered when establishing the fractured medium model. Meanwhile, previous numerical simulation studies have only analyzed one wellbore layout method or compared the heat transfer efficiency of two wellbore layout methods, so it is difficult to analyze the advantages and disadvantages of multiple wellbore layout methods in geothermal energy production. Furthermore, HDR geothermal energy development includes two parts: reservoir reconstruction and thermal energy exploitation. Most previous studies only focus on one part. To restore the entire geothermal energy development process, the two parts should be combined to conduct research.

This paper takes geothermal field data from Matouying as a geological reference. Based on a numerical simulation performed with TOUGH2-BIOT, a random feature to distinguish the direction of the distribution of fractures during model creation is introduced, and based on the permeability tensor method, the porous medium is replaced by the fractured medium. The evolution characteristics of the seepage field during the reservoir stimulation process were analyzed to evaluate the fracturing effect. The wellbore layout position was set according to the reservoir reconstruction result, with the heat transfer efficiency of conventional double vertical wells, horizontal wells, and the double-pipe heat exchange system compared and analyzed. The simulation result was verified by analyzing the temperature field evolution characteristics of the geothermal reservoir after heat transfer.

2. Geothermal Data in the Matouying Geothermal Field

The Matouying uplift area is located in the eastern part of Hebei Province and is a low plain area with large tectonic subsidence, with the deep and large fractures in the field well developed [23]. The specific geographical location of the Matouying geothermal field is shown in Figure 1. The north side of the geothermal field is bounded by the Mabei fault, the west side is bounded by the Baigezhuang fault zone, and the south side is bounded by the Hongfangzhi fault zone [24].

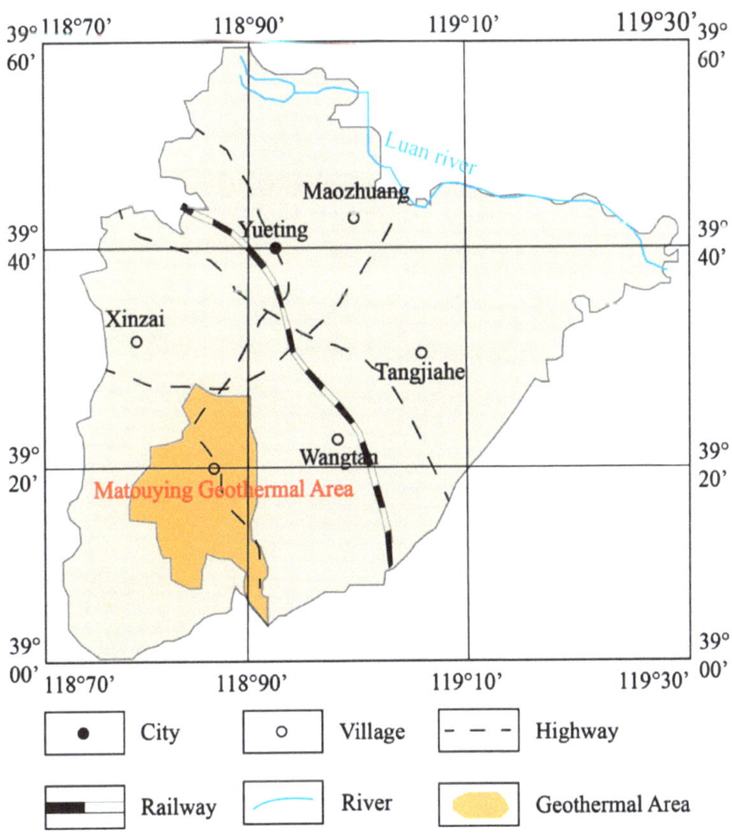

Figure 1. Matouying geothermal field location.

To investigate the deep high-temperature geothermal resource, multiple exploration wells were drilled in the Matouying geothermal field. The temperature logs of exploration wells with the lithologic feature are shown in Figure 2. The measured results indicate that there is a certain degree of heterogeneity in the geothermal gradient in the area, which generally shows that the change is relatively slow in the north and relatively large in the south offshore area. The Cenozoic geothermal gradient in the study area is 3.0–5.0 °C/hm, and the deep geothermal gradient is 3.2 °C/hm, which actually reflects the influence of thermal conductivity on geothermal gradient, as the thermal conductivity of granite and granulite is obviously higher than that of sandstone, mudstone, and loose sediment, generally. In fact, this phenomenon can also be observed in the drilling data of other geothermal fields, such as the Guide EGS field in Qingha, China, and the Feton Hill EGS field in New Mexico, USA. The maximum temperature 4000 m underground can reach 200 °C and the area where the temperature exceeds 150 °C exceeds 600 km^2 in the study area. The reservoir lithology of the geothermal area from top to bottom is Quaternary loose sediment, Neogene mudstone–sandstone mixture, and Taikoo Temple formation metamorphic monzonitic granite and biotite plagioclase granulite.

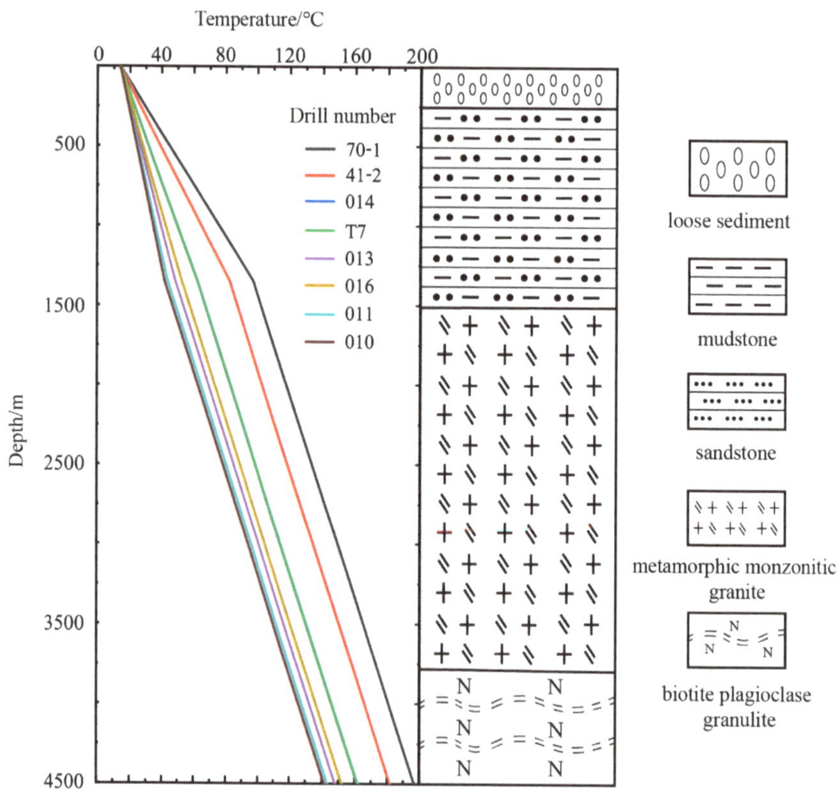

Figure 2. Distributions of rock lithologic and temperature in the depth profile of the borehole in the Matouying uplift area.

From the XRMI televiewer survey in Figure 3, it can be seen that there are a large number of natural fractures at 4200–4500 m underground in the study area, with the density of the natural fracture 0.02–0.26 per/m, and the fracture width in the order of tens of microns. The direction of the maximum principal stress of the formation is SEE-NWW.

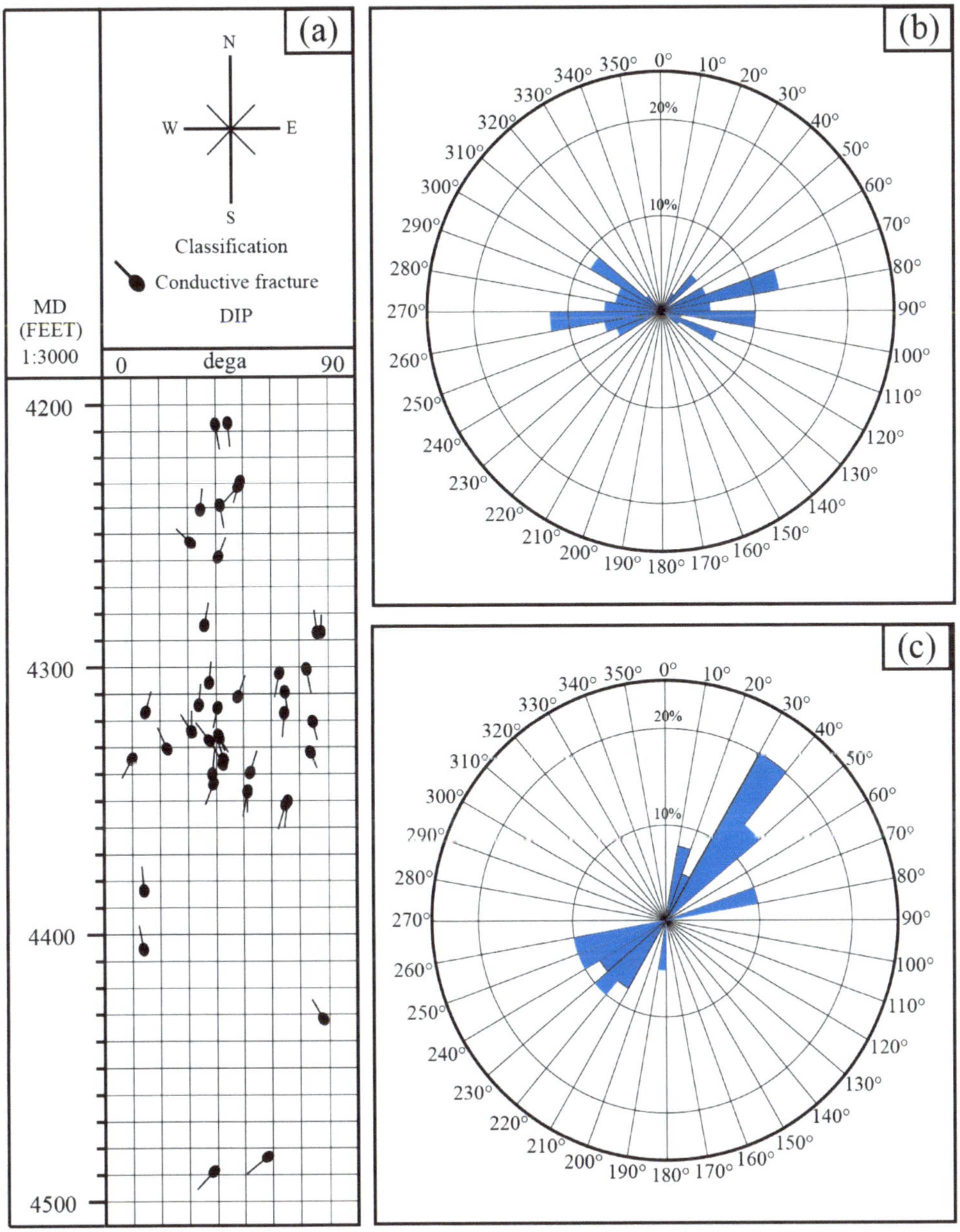

Figure 3. Natural occurring fractures in well M-2: (**a**) tadpole diagram of identified fractures, (**b**) fracture population dip azimuths, and (**c**) fracture population dip strikes.

3. Numerical Models and Simulation Approach

3.1. Governing Equation for THM Process

The coupled THM analysis was conducted using TOUGH2-BIOT [25]. TOUGH2-BIOT does not require external data exchange; therefore, computing ability is more efficient. In addition, TOUGH2-BIOT can be easily extended for considering the THM coupled processes in fractured media [26]. In fact, TOUGHBIOT is an extended version for mechanical problems developed based on TOUGH2 V2.0, which is a well-accepted numerical simulator for solving multiphase fluid and heat flows, both in porous and fractured geological media [27]. The general formulations for multiphase flow, heat convection, and conduction processes are summarized in Table 1.

Table 1. General mathematical model of coupled TH processes in TOUGH2 V2.0.

Description	Governing Equation
Mass and energy conservation	$\frac{d}{dt}\int_V M^\kappa dV = \int_\Gamma F^\kappa \bullet n d\Gamma + \int_V q^\kappa dV$
Mass accumulation	$M^\kappa = \sum_{\beta=A,G} \varphi S_\beta \rho_\beta X_\beta^\kappa, K = w, i, g$
Mass flux	$F^\kappa = \sum_{\beta=A,G} -k\frac{k_{r\beta}\rho_\beta}{\mu_\beta} X_\beta^\kappa \left(\nabla P_\beta - \rho_\beta g\right)$
Energy accumulation	$M^\theta = (1-\varphi)\rho_R C_R T + \sum_{\beta=A,G} \varphi S_\beta \rho_\beta u_\beta$
Heat flux	$F^\theta = -\lambda \nabla T + \sum_{\beta=A,G} h_\beta F_\beta$

Where M^K is the mass accumulation of component κ (kg/m^3), V is the volume (m^3), Γ is the surface area (m^2), M^K is the mass flux of component κ (kg/m^3·s), q^K is the sink/source of component κ (kg/m^3), β = A, G are the aqueous and gaseous phase, respectively, φ is the porosity, S_β is the saturation of phase β, ρ_β is the density of phase β (kg/m^3), X_β^K is the mass fraction of component κ in phase β, k is the permeability (m^2), $k_{r\beta}$ is the relative permeability of phase β, μ_β is the internal energy of phase β (J/kg), P_β is the pressure of phase β (Pa), g is the gravitational acceleration vector (m/s^2), M^θ is the energy accumulation (kg/m^3), ρ_R is the density of rock grain (kg/m^3), C_R is the specific heat of rock grain (J/kg·°C), T is the temperature (°C), μ_β is the viscosity of phase β (Pa·s), F^θ is the energy flux (W/m^2), λ is the average thermal conductivity (W/ °C·m), and h_β is the specific enthalpy of phase β (J/kg·°C).

In mechanics, assuming rock as an elastic material and obeying the generalized Hooke's law, then, based on the stress equilibrium equations, compatibility equations, and stress–strain relationships, the Biot consolidation model can be obtained by combining the effective stress law with stress and displacement as the main unknown variables [28], as shown in Table 2.

Table 2. Three-dimensional extended Biot mechanical model.

Description	Governing Equations
Displacement	$-G\nabla^2 w - \frac{G}{1-2v}\nabla(\nabla \cdot w) + \nabla P_a + 3\beta_T K \nabla T + F = 0$
Normal strain	$\varepsilon = \nabla w$
Effective stress	$\sigma' = \sigma - P = 2G(\frac{v}{1-2v}\varepsilon_V + \varepsilon) + 3\beta_T K \nabla T$

Where G is the shear modulus (Pa), w is the displacement (m), β_T is the thermal expansion coefficient (1/°C), ε_V is the bulk strain, σ is the normal stress (Pa), and ε is the normal strain.

3.2. Permeability Evolution of the Fractured Rock

In this study, the establishment of fractured reservoirs is mainly based on borehole imaging results; therefore, there are many imaginary fractures in each grid in the model. The permeability of fractures with different dip angles and dip tendencies are converted into a permeability tensor to represent the equivalent permeability of a group of fractures with the same dip angle and tendency, which is defined by Equation (1) [29].

$$k = \sum_{i=1}^{n} \frac{b_i^3}{12l_i} \begin{bmatrix} 1 - \cos^2\beta\sin^2\gamma & -\sin\beta\sin^2\gamma\cos\beta & -\cos\beta\sin\gamma\cos\gamma \\ -\sin\beta\cos\beta\sin^2\gamma & 1 - \sin^2\beta\sin^2\gamma & -\sin\beta\sin\gamma\cos\gamma \\ -\cos\beta\sin\gamma\cos\gamma & -\sin\beta\sin\gamma\cos\gamma & 1 - \cos^2\gamma \end{bmatrix} \quad (1)$$

where k is the permeability (μm^2), b is the equivalent hydraulic fracture aperture (m), l is the distance between the fluid flow in the fracture (m), and β and γ are the fracture surface tendency and dip angles, respectively.

The stress state of the fracture surface determines its ability to generate shear failure. After calculating the stress state of each grid, it needs to be converted to the corresponding fracture surface. According to the mechanical balance method, the stress acting on a given fracture plane is calculated [30]. The stress has the following relationship:

$$\sigma'_n = \sigma'_x l^2 + \sigma'_y m^2 + \sigma'_z n^2 \quad (2)$$

$$\tau_n = \left[\left(\sigma'_x - \sigma'_y\right)^2 l^2 m^2 + \left(\sigma'_y - \sigma'_z\right)^2 m^2 n^2 + \left(\sigma'_z - \sigma'_x\right)^2 l^2 n^2 \right]^{1/2} \quad (3)$$

where σ'_n is the effective normal stress (Pa), τ_n is the shear stress (Pa), σ'_i is the effective normal stress in x, y, z directions (Pa), and l, m, n are the direction cosines of the fracture plane normal with respect to the principal stress axes σ'_x, σ'_y, σ'_n, respectively.

The Mohr–Coulomb failure criterion is used to define the shear strength of fractured rock, as follows [31]:

$$F_c = |\tau_n| - \mu_s \sigma'_n - c \quad (4)$$

where F_c is the shear strength (Pa), μ_s is the Static friction coefficient, and c is the cohesion (Pa).

The shear displacement of fracture can be calculated from the stiffness coefficient and the excess shear stress [32], and is written as:

$$d = \frac{\tau_{ex}}{K_f}, \begin{cases} \tau_{ex} = 0 & for F_c < 0 \\ \tau_{ex} = |\tau_n| - \mu_d \sigma'_n & for F_c \geq 0 \end{cases} \quad (5)$$

where d is the shear displacement (m), τ_{ex} is the excess shear stress (Pa), K_f is the shear fracture stiffness (Pa/m), and μ_d is the dynamic friction coefficient.

The rock will experience the change of aperture, bending, and surface roughness after the shear failure. Meanwhile, under the action of overburden load, the rock will produce shear displacement, with the relationship between fracture permeability and related parameters shown in Equation (6).

$$\Delta K = \frac{\Delta K_{\max}}{1 + \exp\left[\log_e(19) \cdot \left(1 - 2.0\frac{d - d_5}{d_{95} - d_5}\right)\right]} \quad (6)$$

where $K = \log_{10}(k)$, $\Delta K_{\max} = \log_{10}(k_{\max}) - \log_{10}(k_{imi})$ is the maximum permeability enhancement, and d5 and d95 are the shear displacements corresponding to 95% and 5% of the maximum permeability enhancement, respectively.

4. Numerical Simulation of Reservoir Stimulation

4.1. Model Description

4.1.1. Model Geometry and Spatial Discretization

The random fractured medium model was developed based on information provided by borehole imaging results. The model geometry is 300 m × 300 m in the horizontal, with the top and bottom surfaces of the model 4200 m and 4500 m underground, respectively. The z-axis of the model coincides with the direction of the maximum principal stress, while the x-axis and y-axis coincide with the horizontal maximum principal stress and minimum principal stress, respectively. When dividing the grid, the center of the injection well is dense in the middle and the surrounding is sparse. The grid size of the central fracturing zone (100 m × 100 m × 100 m) is 10 m, the radius of the water injection well is 0.1 m, and it is located in the center of the fracturing zone; from the fracturing zone to the outside, the grid volume increases to 20 m and 30 m in turn.

4.1.2. Initial and Boundary Conditions

According to the geothermal gradient curve of the study area, the temperature of the reservoir top surface was set to 171 °C. According to the in situ stress formulation, the vertical stress gradient is 26.85 Mpa/km, the horizontal maximum principal stress gradient is 18.80 Mpa/km, and the horizontal minimum principal stress gradient is 16.11 Mpa/km.

The Dirichlet boundary conditions are applied to the injection well. Meanwhile, the frictional pressure loss along the wellbore is ignored, which can effectively conduct the wellhead high pressure to the fracturing point. The zero heat and mass flow boundary conditions are applied to the bottom boundary of the model, and the rest of the boundaries are constant temperature and constant pressure boundaries. To monitor the change of the stress field of the model in real-time, the displacements in the x and y directions of the top, bottom, and surrounding boundaries of the model, and the z-direction displacement of the bottom boundary, are limited.

4.1.3. Initial Permeability

The initial permeability of the fracture media model is based on borehole imaging results. To characterize the difference in fracture density in the vertical direction, the model is divided into five layers from top to bottom, as shown in Table 3. Based on Equation (1), the distribution of initial permeability k_x, k_y, and k_z are calculated and shown in Figure 4. Under the action of the in situ stress field, the permeability in the k_x direction is the best, followed by the k_y direction, and the k_z direction is the smallest, with the k_x direction the optimal direction for reservoir stimulation.

Table 3. Parameter values used for generating random fracture medium model.

Depth (m)	Fracture Density (1/m)	Fracture Tendency			Fracture Dip		Fracture Aperture (mm)
		70°–100°	260°–310°	Other	40°–60°	Other	
4200–4260	0.08	0.28	0.37	0.35	0.5	0.5	0.03379
4260–4300	0.05	0.28	0.37	0.35	0.5	0.5	0.03344
4300–4400	0.26	0.28	0.37	0.35	0.5	0.5	0.03236
4400–4440	0.02	0.28	0.37	0.35	0.5	0.5	0.03164
4440–4500	0.02	0.28	0.37	0.35	0.5	0.5	0.03153

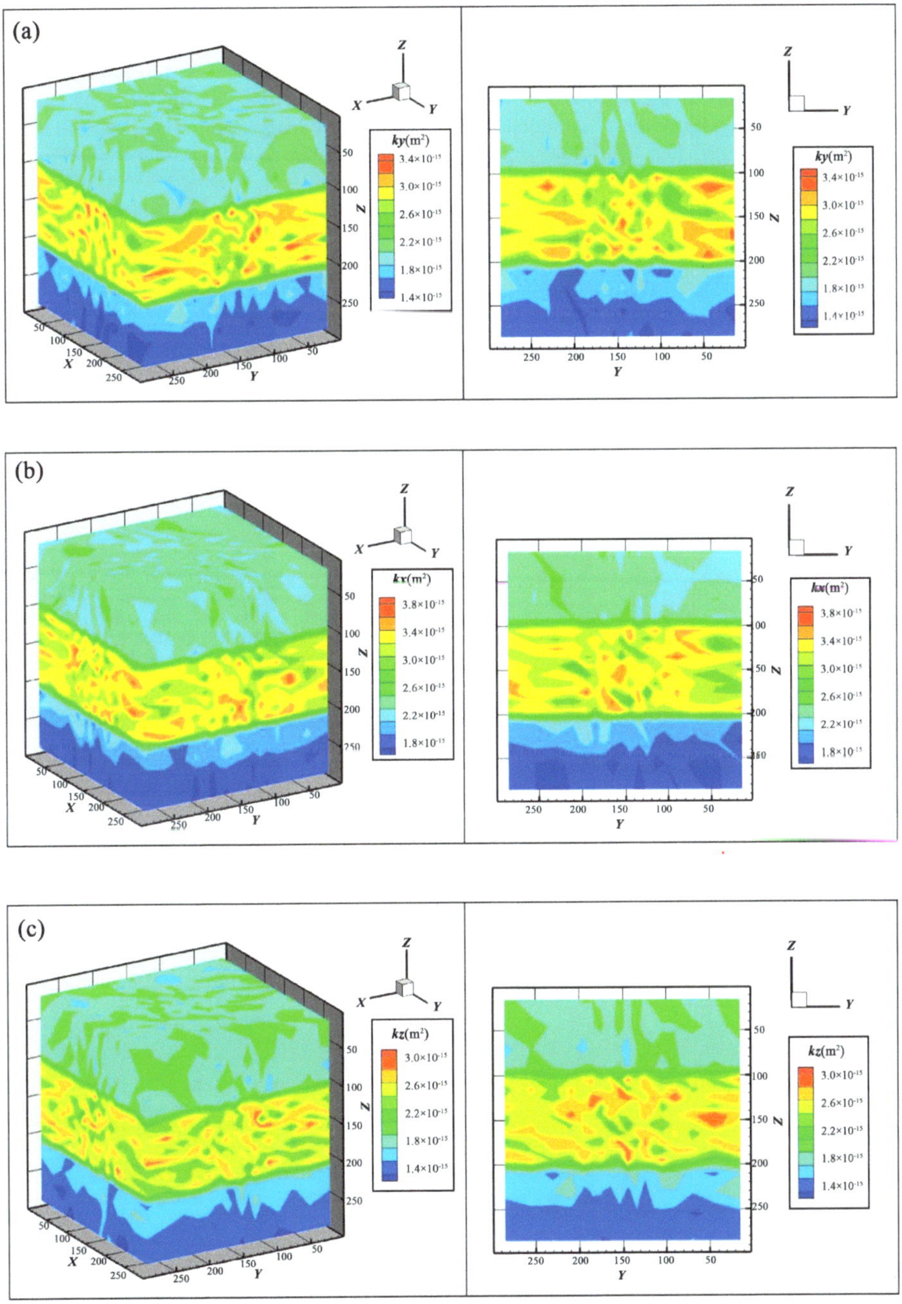

Figure 4. Three-dimensional images display the spatial distribution of initial fracture permeability in the x, y, and z directions, respectively.

4.1.4. Reservoir Parameters

The main parameters of the reservoir are all measured by laboratory experiments, with the static and dynamic friction coefficients chosen concerning the experimental results of [33]. Based on Equations (5) and (6), shear fracture stiffness can be calculated from ΔK_{max}, with the parameter used in Equation (6) referencing the hydro-mechanical experimental results of Lee et al. [34]. The values of the main parameters of the THM model are shown in Table 4.

Table 4. Parameter values used for generating in the THM model.

Material Parameter	Value	Fracture Parameter	Value
Rock density (kg/m^3)	2690	Static friction coefficient	0.65
Porosity	1.6%	Dynamic friction coefficient	0.55
Thermal conductivity (W/m °C)	2.996	Cohesion (MPa)	10.2
Specific heat capacity (J/kg·°C)	946	Shear fracture stiffness (MPa/m)	500
Young's modulus (GPa)	19.95	Permeability Parameter	Value
Poisson's ratio	0.295	ΔK_{max}	1.7
		d_5(mm)	1.5
		d_{95}(mm)	5

The temperature and pressure at the injection point location were estimated by the wellbore simulator T2WELL. As shown in Figure 5, the simulation results show that the temperature of the injected water fluid is stable at 35 °C after 1 day, and the temperature decreases slightly with the increase of the water injection time, but the decrease is small. The pressure in the well is defined as $P_{bottom} = \rho g h + P_{top}$ and the hydrostatic pressure at the bottom of the well is 41 MPa.

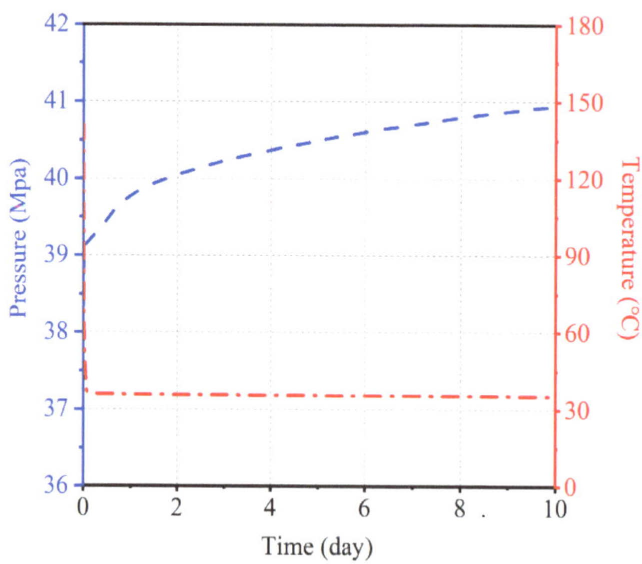

Figure 5. Initial temperature and pressure values at the water injection point.

4.2. Model Calibration

In the random fractured media model, the generation of each fracture is random. In order to reduce this uncertainty, the dip tendency and dip angle of the fractures should not be changed as much as possible during model calibration, and the aperture of the fractures should be adjusted first. In the model calibration stage, the measured injection flow rate is used to simulate the changing trend of wellhead pressure during the reservoir reconstruction process, with the measured data and simulation results fitted by adjusting the fracture width. The adjustment range of each parameter was determined based on the relevant research results, to prevent excessive adjustment.

Reservoir fracturing mainly includes three stages; in the first stage, the water injection rate is 2.0–3.0 m^3/min, which lasts for 2 days. As shown in Figure 6, the simulated curve is consistent with the measured head pressure curve. Within the first 7 h of the first day, the wellhead pressure was relatively stable, indicating at this stage that the hydraulic aperture of the conductive fractures in the reservoir gradually increased with the increase in water injection. However, in the eighth hour of the injection, the wellhead pressure showed an upward trend, and this trend continued to the end of the second day. It indicates that the hydraulic aperture of conductive fractures in the reservoir reaches the maximum under the action of in situ stress, but no new fractures are generated.

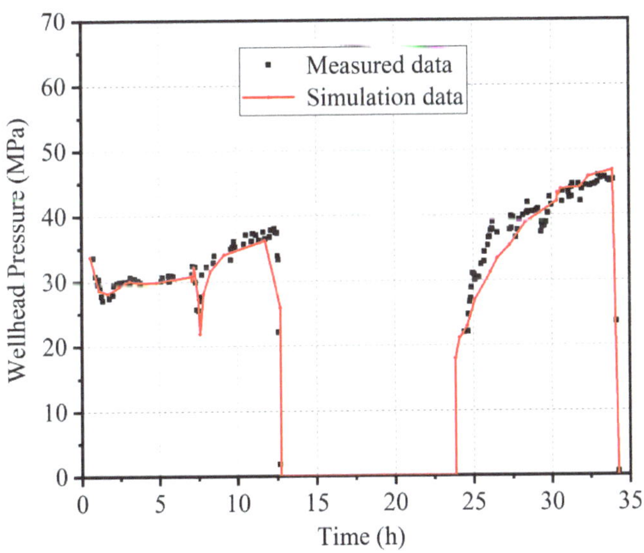

Figure 6. Comparison of measured and simulated results of reservoir stimulation in the first stage.

In the second stage, the water injection rate is 4.0–5.5 m^3/min, which lasts for 3 days. As shown in Figure 7, on the third and fourth days of fracturing and the first half of the fifth day, although the curve fluctuates, the wellhead pressure does not exceed the first stage, indicating that shear failure did not occur in this stage. The reason for this phenomenon is that the liquid pressure inside the fracture relieves the effect of in situ stress so that the non-conducting fractures in the reservoir become a conducting fracture. This trend is alleviated after the fracturing time is 100 h, and the wellhead pressure further increases.

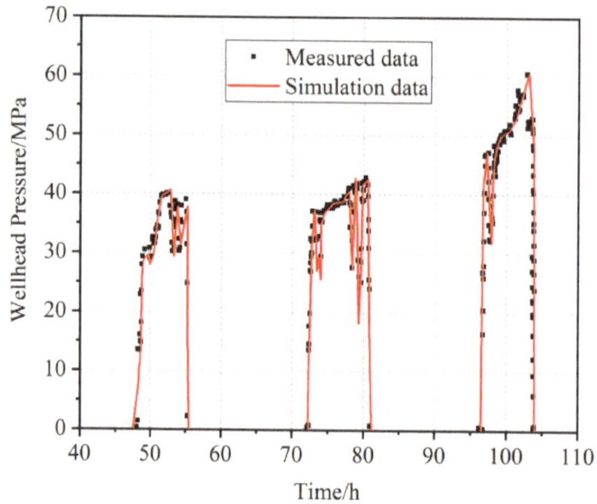

Figure 7. Comparison of measured and simulated results of reservoir stimulation in the second stage.

In the third stage, the water injection rate is 6.0–7.0 m^3/min, which lasts for 3 days. As shown in Figure 8, compared with the previous two stages, the wellhead pressure increased significantly in this stage, and compared with the second stage, the wellhead fluctuates more sharply, indicating that the hydraulic shear phenomenon occurred in this stage. On the seventh and eighth days of fracturing, the maximum value of wellhead pressure gradually decreased. The reason for this phenomenon is that due to the occurrence of hydraulic shear, the original fractures and new fractures of the reservoir are connected, and an artificial fracture network is gradually formed.

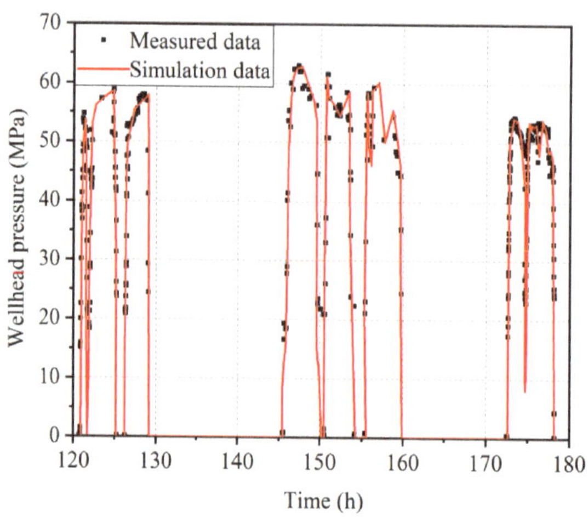

Figure 8. Comparison of measured and simulated results of reservoir stimulation in the third stage.

4.3. Reservoir Reconstruction Results

The basic principle of reservoir reconstruction is to increase the liquid pressure in the reservoir or reduce the temperature of the reservoir, thereby reducing the effective stress of the rock body firmware, causing the rock to undergo shear slip, and thereby increasing its permeability. This can be represented by a Mohr diagram and the movement of stress circles to demonstrate such a physical process, as shown in Figure 9.

After the first stage, the Mohr stress circle expands outward from the center of the circle. At this time, the Mohr stress circle is far from the failure envelope and no fractures are formed in the reservoir (Figure 9a). After the second stage, the center of the Mohr stress circle moves to the left and the diameter of the Mohr circle increases significantly. At this time, the distance between the Mohr circle and the failure envelope line is small (Figure 9b). After the third stage, the center of the Mohr stress circle continues to move to the left and, at the same time, its diameter further increases, and the Mohr stress circle intersects with the failure envelope. At this time, some rocks near the fracturing point suffered shear damage (Figure 9c).

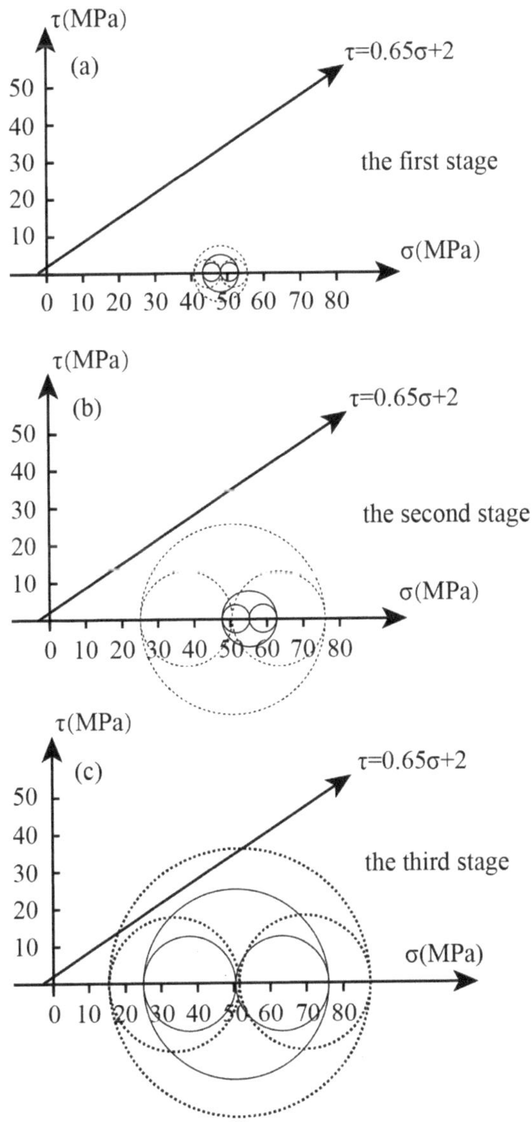

Figure 9. Stress state of natural fracture around the fracturing point after each stage: (**a**) first stage; (**b**) second stage; (**c**) third stage.

The main purpose of reservoir reconstruction is to increase the connectivity of artificial fracture networks, that is, the permeability of fluid media in the fractures. Therefore, to reflect the reformation effect of this reservoir reconstruction, the permeability of the reservoir after the end of each stage was analyzed.

After the first stage, the permeability of the reservoir is enhanced in all directions. As shown in Figure 10, the enhancement effect of permeability along the x direction is the best, the enhancement range is 25 m, and the reservoir permeability at the water injection point is enhanced by 17 times; the enhancement range of permeability in the y direction is 18 m, and the reservoir permeability at the water injection point is enhanced by 7 times; the enhancement range of permeability in the z direction is 23 m, the reservoir permeability at the water injection point is enhanced by 16 times, and the volume of the enhanced permeability area is approximately 10,000 m^3.

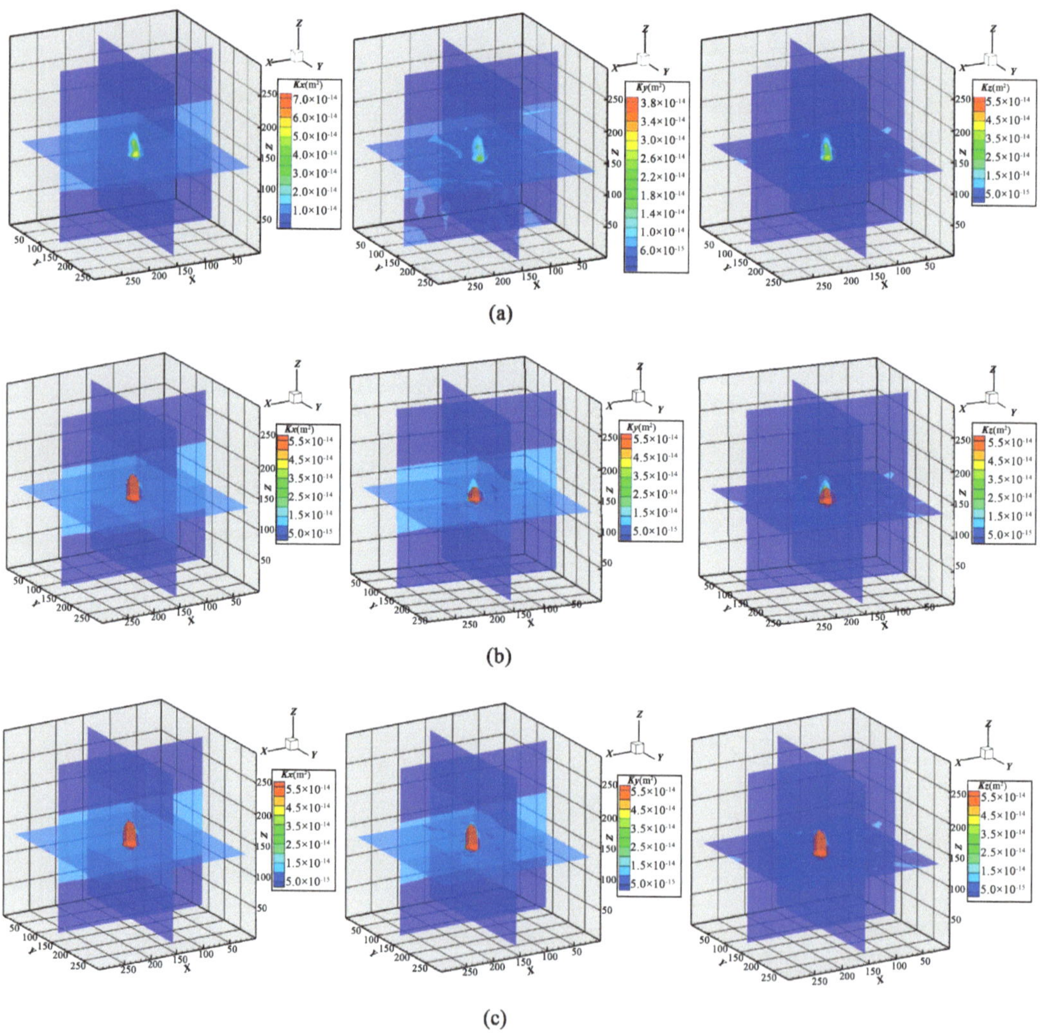

Figure 10. Permeability of reservoir in x, y, z directions after: (**a**) first stage; (**b**) second stage; (**c**) third stage.

As shown in Figure 11, after the second stage, the enhancement range of the permeability in the x direction is 30 m, and the reservoir permeability at the water injection point is enhanced by 256 times; the enhancement range of the permeability in the y direction is 25 m, and the reservoir permeability at the water injection point is enhanced by 121 times; the enhancement range of the permeability in the z direction is 29 m, the reservoir permeability at the water injection point is enhanced by 221 times, and the volume of the enhanced permeability area is approximately 21,000 m^3.

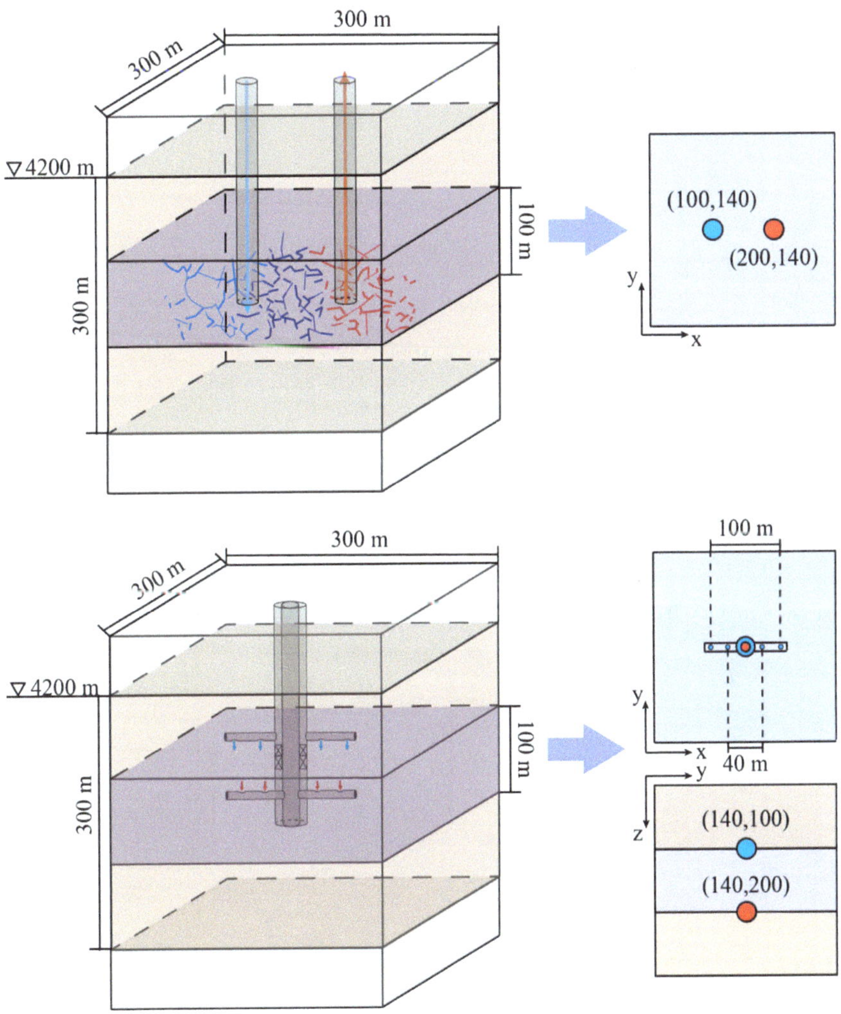

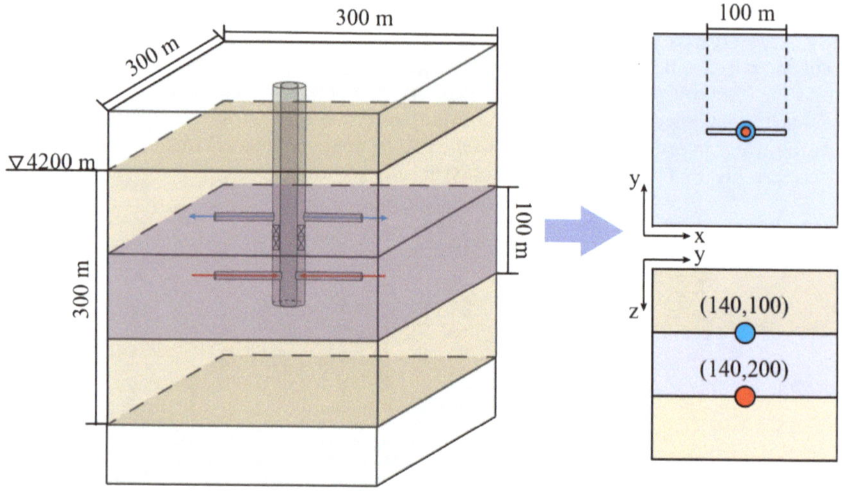

Figure 11. Physical models of the conventional double vertical wells model, horizontal wells, and double-pipe heat exchange system.

As shown in Figure 12, after the third stage, the enhancement range of the permeability in the x direction is 34 m, and the reservoir permeability at the water injection point is enhanced by 818 times; the enhancement range of the permeability in the y direction is 30 m, and the reservoir permeability at the water injection point is enhanced by 312 times; the enhanced range of permeability in the z direction is 33 m, the reservoir permeability at the water injection point is enhanced by 777 times, and the volume of the enhanced permeability area is approximately 33,000 m^3.

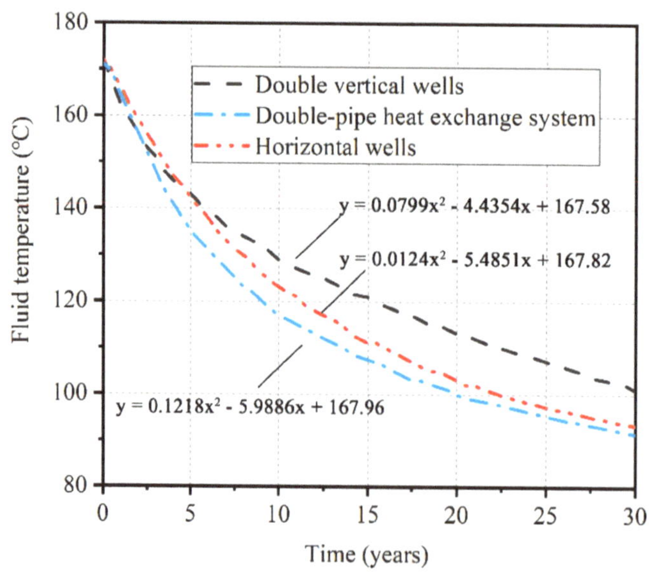

Figure 12. Average production temperature curve.

4.4. Heat Exploitation Model Establish

Based on the results of reservoir stimulation, the conventional double vertical wells, horizontal wells, and double-pipe heat exchange system model were established and are shown in Figure 11. In order to prevent water leakage, according to the actual conditions of the reservoir, a layer of dense rock is set on the upper and lower parts of the model. The total water injection rate during the simulation was 15 kg/s and the total heat exploitation period is 30 years.

To facilitate the analyses and discussion, three parameters, namely, average production temperature, output thermal power, and heat extraction rate, are defined to characterize the heat exploitation performance of the EGS.

The average production temperature is defined as follows [20]:

$$T_{pro} = \frac{\int_S T(t)dS}{S} \tag{7}$$

where S is the perimeter of the production well (m) and $T(t)$ is the temperature of the production well at time t (°C).

The output thermal power is calculated as follows:

$$P = q\rho_W c_W \left(T_{pro} - T_{inj}\right) \tag{8}$$

where q is the average volume flow rate of the production well (m^3/s), T_{inj} is the temperature of injection well (°C), P_w is the density of water (kg/m^3), and c_w is the heat capacity of water (J/kg·°C).

The heat extraction rate is equal to the heat extracted divided by the heat energy stored in the simulated reservoir volume, written as:

$$\eta = \frac{\iint_\Omega \rho_r c_r [T_{ini} - T(t)]d\Omega}{\iint_\Omega \rho_r c_r (T_{ini} - T_{inj})d\Omega} \tag{9}$$

where Ω is the simulated reservoir volume (m^3).

5. Simulation Results and Analysis

5.1. Comparison of Heat Exploitation Performance

The simulation results based on the heat exploitation model are shown in Figures 12–14; to further compare the heat extraction capabilities of different systems, the trend line equations in polynomial form were also included in the figure. Under the same reservoir condition, Figure 12 shows the temperature of a production well under different production models over 30 years. It can be observed that the average production temperature of the three exploitation models decreases gradually with time. After 30 years, the temperature of a production well in the conventional double vertical wells model, horizontal wells, and double-pipe heat exchange system is 101 °C, 93.4 °C, and 91.6 °C, a decrease of 41.2%, 45.7%, and 46.7%, respectively. In the process of geothermal energy exploitation, the outlet temperature of the double-pipe heat exchange system is always lower than the horizontal wells. After 2 years of exploitation, the outlet temperature of the conventional double vertical wells model is higher than the double-pipe heat exchange system and exceeds the horizontal well in the fifth year.

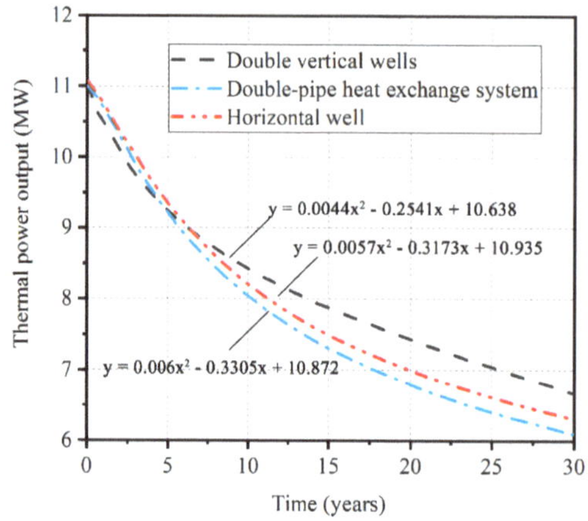

Figure 13. Thermal power output curve.

As shown in Figure 13, the thermal power output curve is highly similar to the fluid temperature curve. After 30 years, the thermal power output in the conventional double vertical wells model, horizontal wells, and double-pipe heat exchange system is 6.67 MW, 6.31 MW, and 6.1 MW, a decrease of 39.4%, 42.6%, and 44.5%, respectively. In the 30th year of heat exploitation, the thermal power of conventional double vertical wells is 6% higher than that of horizontal wells and 11% higher than the double-pipe heat exchange systems.

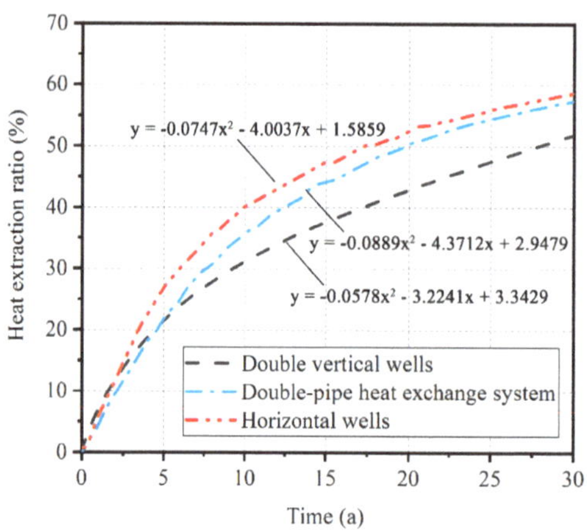

Figure 14. Heat extraction ratio curve.

As shown in Figure 14, after 30 years of geothermal energy exploitation, the heat extraction ratio of the double-pipe heat exchange system is always lower than the horizontal wells. In the early stage of geothermal energy exploitation, the heat transfer efficiency of the conventional double vertical wells is higher, but as the exploitation progresses, it is gradually lower than that of the horizontal wells and the double-pipe heat exchange system. Additionally, in the 30th year of heat exploitation, the heat extraction ratio of the horizontal wells is 2% higher than the double-pipe heat exchange system and 6.5% higher than the conventional double vertical wells.

5.2. Analysis of the Temperature Field Results

The temperature field distribution under different exploitation models is shown in Figures 15–17. In the early stage of geothermal energy exploitation, the conventional double vertical wells are centered on the injection point, the low-temperature area is approximately concentric circles, and the range of the low-temperature area is small and does not extend to the vicinity of the production well. Therefore, compared with other production models, the heat extraction rate of the double vertical wells system is higher at this stage. With the development of geothermal energy exploitation, the fluid medium flows continuously to the outlet point through the fracture network, the low-temperature area expanded to the outlet point. In the 20th year, the low-temperature area was extended to the water outlet point, the low-temperature area is in the shape of a droplet, and at this stage, the production efficiency of geothermal wells decreased significantly. In the 30th year, the low-temperature area spreads more below than above the water outlet. It is speculated that the reason for this phenomenon is that the fluid medium will flow downward due to the action of gravity. At the same time, compared with other production modes, the conventional double vertical production mode has only one water outlet. This will lead to untimely water output from the outlet point, and the sedimentary fluid medium continuously exchanges heat with the reservoir, which is also the reason why the water temperature and thermal power output of the double vertical wells are higher than in other heat extraction systems.

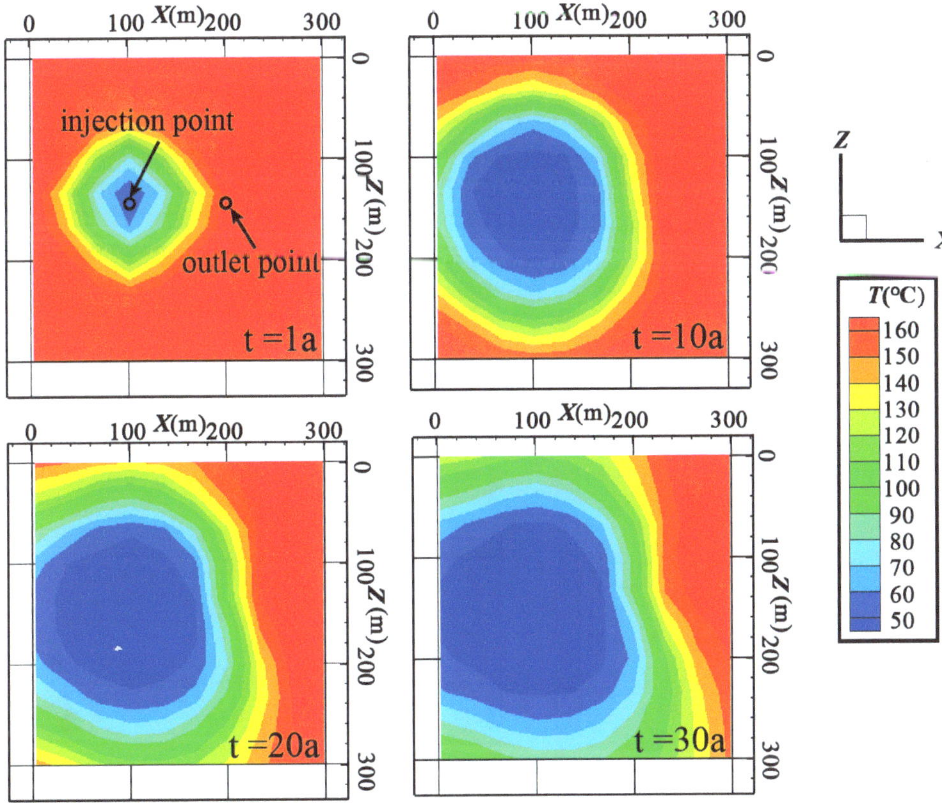

Figure 15. Temperature distribution for the conventional double vertical wells.

In the early stage of geothermal energy exploitation, the temperature field of the horizontal wells model is an irregular ellipsoid with the well as the axis, and the closer to the direction of the x-axis, the larger the area of the low-temperature area. This is because the x direction is the dominant direction of the fluid medium. With the development of geothermal energy exploitation, the low-temperature area gradually extends to the production well, and the boundary of the low-temperature area near the water outlet is linear. This shows that under the condition of the same injection rate, the fluid medium in the horizontal well production system can be discharged from the outlet point in time. In addition, after 30 years, the leading edge of the low-temperature area reaches the top of the outlet point; however, the temperature of the reservoir below the outlet point is still high, so the production well still produces high-temperature fluids. Compared with traditional double vertical wells, the low-temperature area of horizontal wells is larger, which means that it has a greater heat extraction ratio.

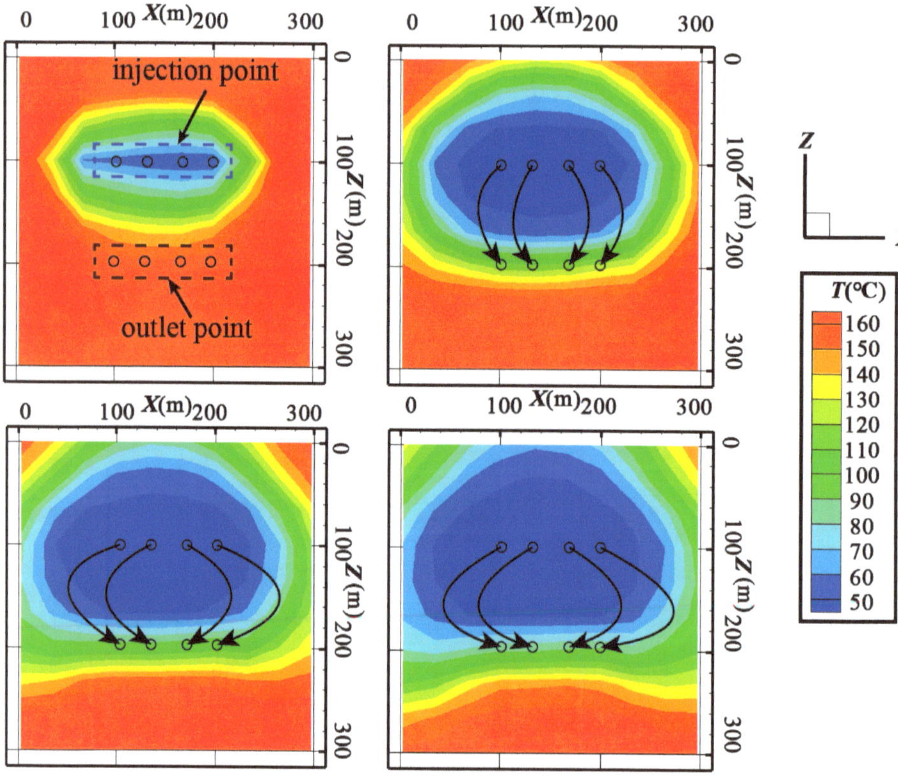

Figure 16. Temperature distribution for the horizontal wells.

The geothermal field characteristics of the double-pipe heat exchange system are similar to those of the horizontal wells, expanding to the outside with the injection well as the center. In contrast, the leading edge of the low-temperature area is wavy; this situation is similar to conventional double vertical wells. Compared with horizontal wells, the expansion area of the low-temperature area of the double-pipe heat exchange system is smaller, which indicates that its heat extraction efficiency is lower than that of horizontal wells. However, in this exploitation model, the temperature reduction of the reservoir is small, which is conducive to the continuous exploitation of geothermal energy.

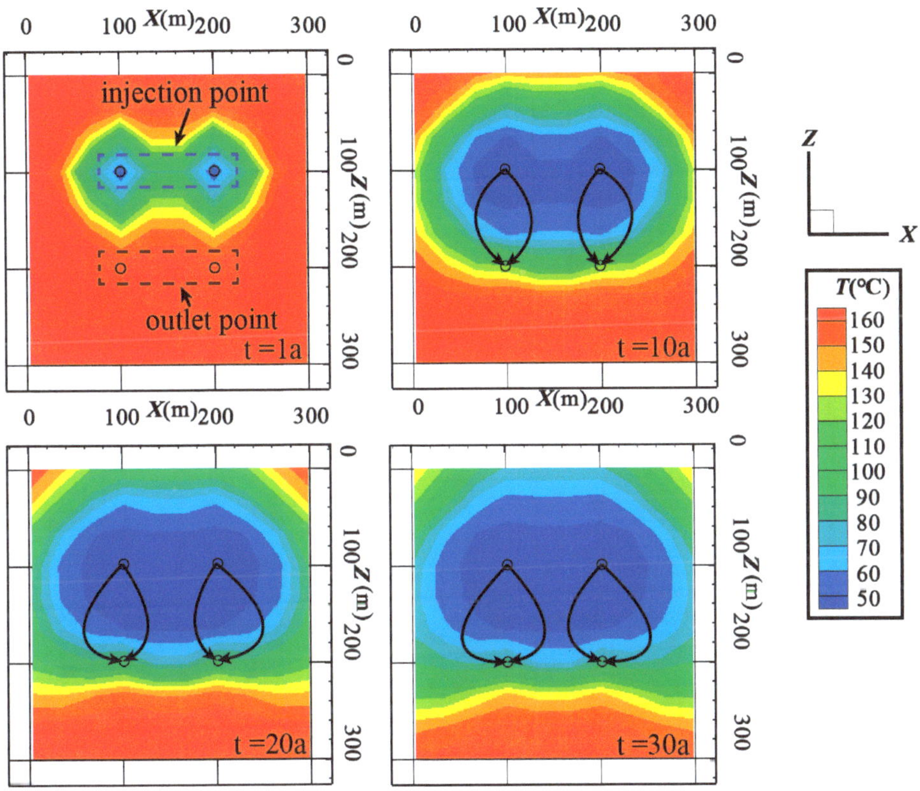

Figure 17. Temperature distribution for the double-pipe heat exchange system.

5.3. Analysis of the Pressure Field Results

The pressure field distribution under different exploitation models is shown in Figures 18–20. As shown in Figure 18, in the early stage of geothermal energy exploitation, it can be observed that the pressure around the injection point is the highest, while the pressure around the outlet point is the lowest. This is because the conventional double vertical wells use single-point water injection and single-point pumping. Compared with other heat-transfer modes, the seepage path of the fluid medium in the fracture network is longer and the loss of wellhead pressure is larger. In this stage, the low-temperature area did not extend to the vicinity of the outlet point; as the water injection continues, the pores and fractures in the reservoir are filled with a fluid medium, and the reservoir pressure gradually increases with time. After 20 years of exploitation, the injection pressure increases and the production pressure decreases. In this stage, the low-temperature area extends to the vicinity of the outlet point. As the temperature of the reservoir decreases, the viscosity of the fluid medium increases, which leads to the wellhead pressure lost during the flow of the fluid increases, and the pressure difference between the water injection point and the outlet point increases.

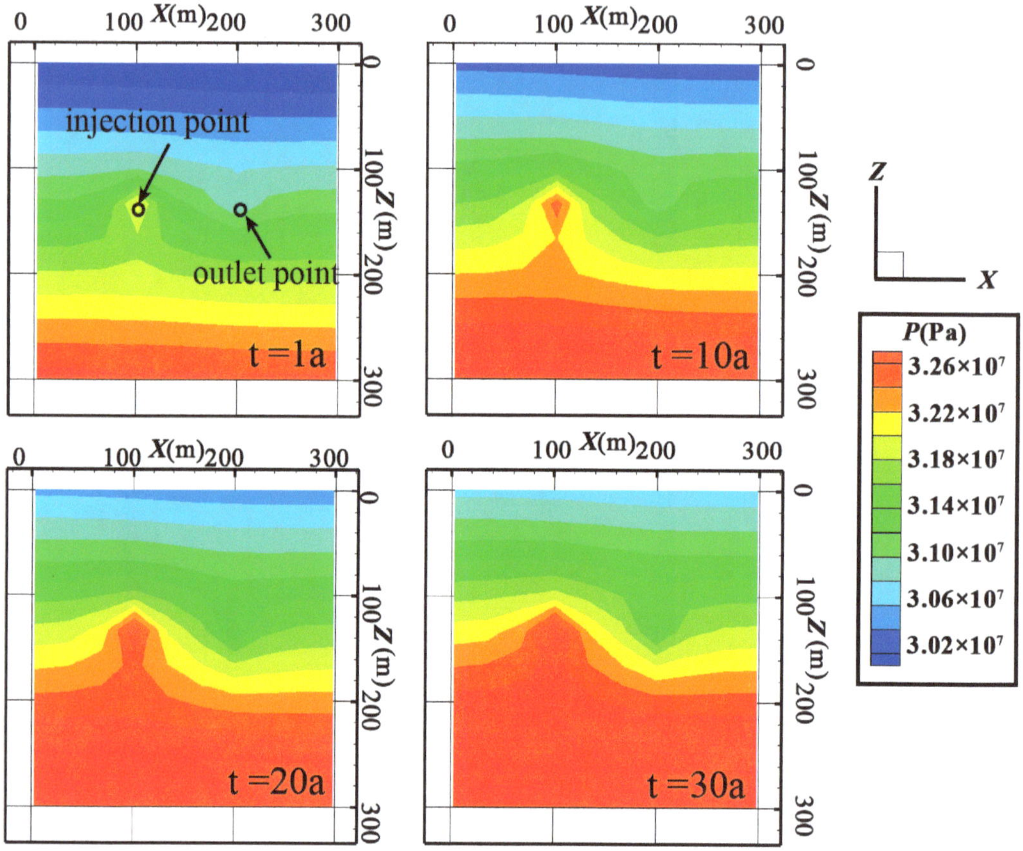

Figure 18. Pressure distribution for the conventional double vertical wells.

As shown in Figure 19, compared with the conventional double vertical wells, after 30 years of horizontal wells exploitation, there is very little change in reservoir pressure. This is because there are many inlet points and outlet points in the horizontal wells; under the condition of the same total injection rate, the injection volume at each water inlet point is relatively small, resulting in a decrease in the water pressure at the inlet point. At the same time, when the water is pumped, the fluid medium can be continuously and efficiently pumped from the output point, with a small wellhead pressure decrease at the water outlet point. Therefore, the overall pressure of the reservoir increases with the increase of exploitation time. However, the dispersed water flow will take away more heat from the geothermal reservoir, which also leads to the expansion of the low-temperature area in the reservoir; during the heat transfer process, the wellhead pressure loss of the fluid medium increases, and there is an appearance of a low-pressure area near the outlet point.

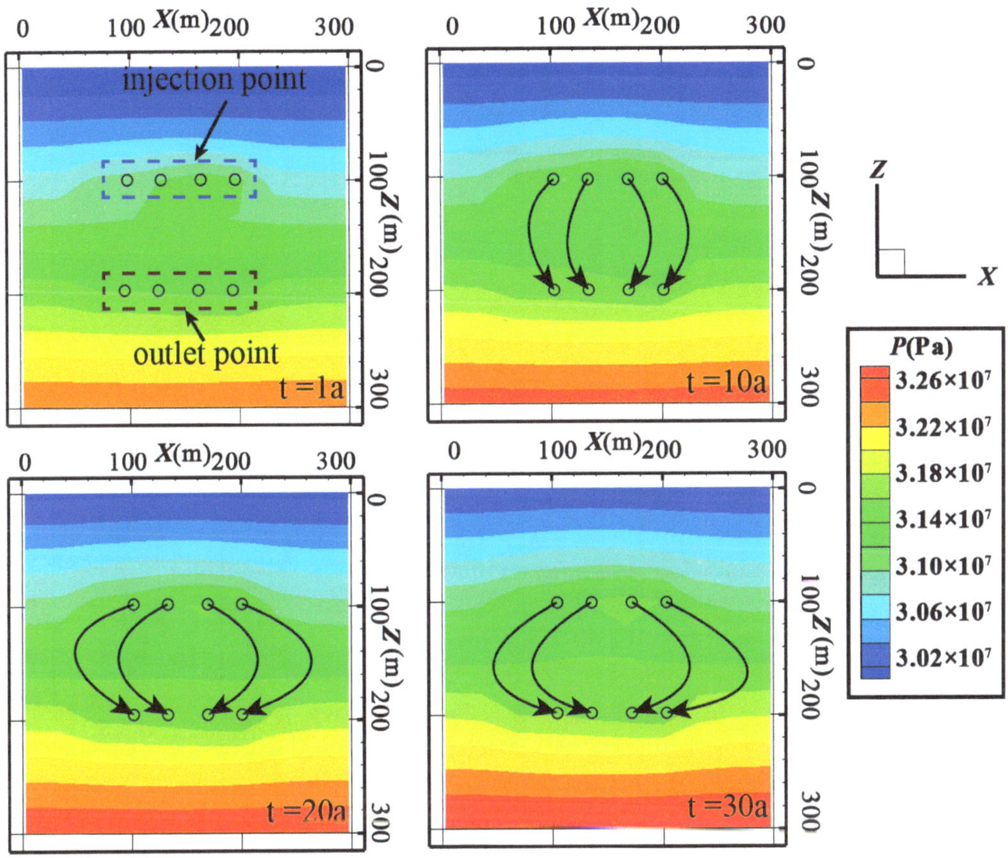

Figure 19. Pressure distribution for the horizontal wells.

As shown in Figure 20, compared to other thermal exploitation models, the change near the injection point and the outlet point is the most obvious in the double-pipe heat exchange system. This is because the number of injection points and outlet points is intermediate between the other two thermal exploitation modes. In the long-term heat generation process of up to 30 years, injection and production of heat exchange fluid will not lead to a significant or large-range increase in reservoir pressure; meanwhile, it will not cause the reservoir to cool down significantly due to the excessive dispersion of water injection. The internal pressure distribution of the reservoir at 30a is not significantly different from that at 20a. It can be seen that in the late production period, the pressure of the thermal reservoir has risen very slowly, and the reservoir pressure is nearly stable, which will not have a negative impact on the long-term stable operation of the geothermal well. In summary, using the double-pipe heat exchange system for geothermal exploitation has certain advantages.

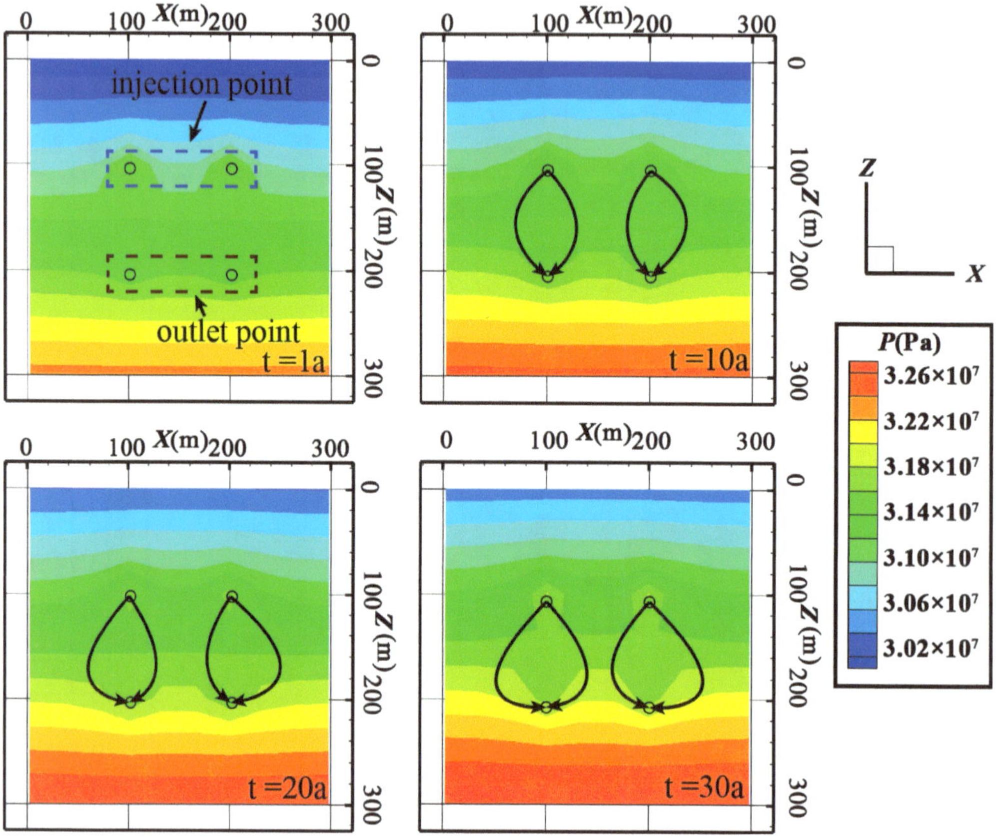

Figure 20. Pressure distribution for the double-pipe heat exchange system.

6. Conclusions

In this paper, using the Matou Camp hot field data as a geological reference, a random fractured media model is established based on TOUGH2-BIOT which analyzes the evolution characteristics of the seepage field during the reservoir stimulation process and evaluates the fracturing effect. According to the results of reservoir stimulation, the wellbore layout position is set, and the heat transfer performance of the conventional double vertical well, horizontal well, and double tube heat exchange system is compared and analyzed. Based on the simulation results, the following conclusions are drawn from this study.

(1) Reservoir reconstruction is mainly divided into three stages: In the first stage, the hydraulic aperture of the conducting fractures reaches the maximum value; in the second stage, the non-conductive fractures overcome the in situ stress and become conducting fractures; in the third stage, the rock in the reservoir undergoes shear failure, fractures expand and connect, and, finally, a fracture network is formed.

(2) After the first stage, the enhancement ranges of k_x, k_y, and k_z are 25, 18, and 23 m, respectively. Compared with the initial water injection point, the permeability at k_x, k_y, and k_z is increased by 17, 18, and 16 times, respectively. After the second stage, the enhancement ranges of k_x, k_y, and k_z are 30, 25, and 29 m, respectively. The permeability at k_x, k_y, and k_z is increased by 256, 121, and 221 times, respectively. After the third stage, the enhancement ranges of k_x, k_y, and k_z are 34, 30, and 33 m,

respectively. The permeability at k_x, k_y, and k_z is increased by 818, 312, and 777 times, respectively. After each stage, the volume of the enhanced permeability area is approximately 10,000, 21,000, and 33,000 m^3, respectively.

(3) In the process of geothermal energy exploitation, the outlet temperature and thermal power output of the double-pipe heat exchange system are always lower than the horizontal wells. After 5 years of exploitation, the outlet temperature and thermal power output of traditional double vertical wells gradually exceeded that of horizontal wells and double-pipe heat exchange systems. After 30 years, the temperature of a production well in the conventional double vertical wells model, horizontal wells, and double-pipe heat exchange system is 101 °C, 93.4 °C, and 91.6 °C, a decrease of 41.2%, 45.7%, and 46.7%, respectively. The thermal power output is 6.67 MW, 6.31 MW, and 6.1 MW, a decrease of 39.4%, 42.6%, and 44.5%, respectively.

(4) The heat extraction ratio of the double-pipe heat exchange system is always lower than the horizontal wells. In the early stage of geothermal energy exploitation, the heat transfer efficiency of the conventional double vertical wells is higher, but as the exploitation progresses, it is gradually lower than that of the horizontal wells and the double-pipe heat exchange system. After 30 years, the heat extraction ratio of the horizontal wells is 2% higher than the double-pipe heat exchange system and 6.5% higher than the conventional double vertical wells.

(5) The conventional double vertical production mode has only one water outlet, which will lead to untimely water output from the outlet point, while the sedimentary fluid medium continuously exchanges heat with the reservoir, which is also the reason why the water temperature and thermal power output of the double vertical wells are higher than other heat extraction systems. There are many water inlet and outlet points in horizontal wells. When the total water injection volume is the same, the dispersed water flow will transfer more heat from the geothermal reservoir, resulting in the expansion of the low-temperature area in the reservoir, which is also the main reason for the higher heat extraction rate to the horizontal wells than other exploitation modes. Compared with other production methods, although the heat exchange efficiency of the double-pipe heat exchange system is lower, the injected fluid will not generate a partial high-pressure area due to the excessive injection, meanwhile, it will not cause the reservoir to cool down significantly due to the excessive dispersion of water injection.

Author Contributions: Conceptualization, J.Z.; Methodology, B.F.; Investigation, H.R.; Writing—review & editing, Z.C.; Project administration, X.L. All authors have read and agreed to the published version of the manuscript.

Funding: This work was performed with the support of the National Key R&D Program of China (No. 2018YFB1501802), S&T Program of Hebei Provincial (No.20374201D), Chinese Academy of Geological Sciences Fund (No.SK202104), Jilin Provincial Department of Science and Technology (No.20200403147SF), Academician Workstation and Academician Cooperation Key Unit Construction of the Second Geological Team of Hebei Coal Geology Bureau (No.205A4101H), and "Research and application demonstration of single well heat recovery technology and ground comprehensive heat utilization technology for dry hot rock thermal energy" (No.19274102D).

Institutional Review Board Statement: Not applicable.

Informed Consent Statement: Not applicable.

Data Availability Statement: The original contributions presented in the study are included in the article. Further inquiries can be directed to the corresponding author.

Acknowledgments: The authors would like to thank the relevant institutions for financial and technical support.

Conflicts of Interest: The authors declare no conflict of interest.

References

1. Moya, D.; Aldás, C.; Kaparaju, P. Geothermal energy: Power plant technology and direct heat applications. *Renew. Sustain. Energy Rev.* **2018**, *94*, 889–901. [CrossRef]
2. Tester, J.W.; Livesay, B.; Anderson, B.J.; Moore, M.C.; Bathchelor, A.S.; Nichols, K.; Blackwell, D.D.; DiPippo, R.; Drake, E.M.; Garnish, J.; et al. *The Future of Geothermal Energy: Impact of Enhanced Geothermal Systems (EGS) on the United States in the 21st Century*; Massachusetts Institute of Technology: Cambridge, MA, USA, 2006.
3. Gupta, H.; Roy, S. *Geothermal Energy: An Alternative Resource for the 21 st Century*; Elsevier: Amsterdam, The Netherlands, 2007.
4. Tomasini-Montenegro, C.; Santoyo-Castelazo, E.; Gujba, H.; Romero, R.J.; Santoyo, E. Life cycle assessment of geothermal power generation technologies: An updated review. *Appl. Therm. Eng.* **2017**, *114*, 1119–1136. [CrossRef]
5. Zhao, X.G.; Wan, G. Current situation and prospect of China's geothermal resources. *Renew. Sustain. Energy Rev.* **2014**, *32*, 651–661. [CrossRef]
6. Jiang, F.; Chen, J.; Huang, W.; Luo, L. A three-dimensional transient model for EGS subsurface thermo-hydraulic process. *Energy* **2014**, *72*, 300–310. [CrossRef]
7. Ogino, F.; Yamamura, M.; Fukuda, T. Heat transfer from hot dry rock to water flowing through a circular fracture. *Geothermics* **1999**, *28*, 21–44. [CrossRef]
8. Wan, Z.; Zhao, Y.; Dong, F.; Feng, Z.J.; Zhang, N.; Wu, J.W. Experimental study on mechanical characteristics of granite under high temperatures and triaxial stresses. *J. Rock Mech. Eng.* **2008**, *27*, 72–77.
9. Klyuev, R.V.; Golik, V.I.; Bosikov, I.I. Comprehensive assessment of hydrogeological conditions for the formation of mineral water resources of the Nizhne-Karmadon deposit. *Geo. Asset. Eng.* **2021**, *332*, 206–218. [CrossRef]
10. Brown, D.W.; Duchane, D.V. Scientific progress on the Fenton Hill HDR project since 1983. *Geothermics* **1999**, *28*, 591–601. [CrossRef]
11. Niibori, H.Y. Numerical models of HDR geothermal reservoirs—A review of current thinking and progress. *Geothermics* **1999**, *28*, 507–518.
12. Rutqvist, J.; Börgesson, L.; Chijimatsu, M.; Kobayashi, A.; Jing, L.; Nguyen, T.S.; Noorishad, J.; Tsang, C.-F. Thermohydromechanics of partially saturated geological media: Governing equations and formulation of four finite element models. *Int. J. Rock Mech. Min. Sci.* **2001**, *38*, 105–127. [CrossRef]
13. Lei, H.W. Characteristics and Mechanism Analysis of Land Subsidence in XX City and Numerical Simulation Research. Ph.D. Thesis, China University of Geosciences, Wuhan, China, 2010.
14. Lu, J.; Ghassemi, A. Coupled Thermo–Hydro–Mechanical–Seismic Modeling of EGS Collab Experiment 1. *Energies* **2021**, *14*, 446. [CrossRef]
15. Mahmoodpour, S.; Singh, M.; Obaje, C.; Tangirala, S.K.; Reinecker, J.; Bär, K.; Sass, I. Hydrothermal Numerical Simulation of Injection Operations at United Downs, Cornwall, UK. *Geoscience* **2022**, *12*, 296. [CrossRef]
16. Saeed, M.; Mrityunjay, S.; Ramin, M.; Sri, K.T.; Kristian, B.; Ingo, S. Direct numerical simulation of thermo-hydro-mechanical processes at Soultz-sous-Forêts. *arXiv* **2022**, arXiv:2206.01830.
17. Saeed, M.; Mrityunjay, S.; Ramin, M.; Sri, K.T.; Kristian, B.; Ingo, S. Impact of Well Placement in the Fractured Geothermal Reservoirs Based on Available Discrete Fractured System. *Geosciences* **2022**, *12*, 19.
18. Gao, K.; Liu, W.; Ma, T.; Hu, Y.; Fang, T.; Ye, L. Numerical Simulation Study of a Novel Horizontally Layered Enhanced Geothermal System: A Case Study of the Qiabuqia Geothermal Area, Qinghai Province, China. *J. Therm. Sci.* **2021**, *30*, 1328–1340. [CrossRef]
19. Cui, G.; Pei, S.; Rui, Z.; Dou, B.; Ning, F.; Wang, J. Whole process analysis of geothermal exploitation and power generation from a depleted high-temperature gas reservoir by recycling CO_2. *Energy* **2021**, *217*, 119340. [CrossRef]
20. Song, X.; Shi, Y.; Li, G.; Yang, R.; Wang, G.; Zheng, R.; Li, J.; Lyu, Z. Numerical simulation of heat extraction performance in enhanced geothermal system with multilateral wells. *Appl. Energy* **2018**, *218*, 325–337. [CrossRef]
21. Mrityunjay, S.; Saeed, M.; Reza, E.; Mohamad, R.S.; Ingo, S. Numerical simulation of a comparative study on heat extraction from Soultz-sous-Forêts geothermal field using supercritical carbon dioxide and water as a working fluid. *arXiv* **2022**, arXiv:2207.07514.
22. Mahmoodpour, S.; Singh, M.; Turan, A.; Bär, K.; Sass, I. Hydro-Thermal Modeling for Geothermal Energy Extraction from Soultz-Sous-forêts, France. *Geosciences* **2021**, *11*, 464. [CrossRef]
23. Cui, L.X.; Wang, L. A Study on Determination of Soil Moisture Content Based on Remote Sensing Images. *J. Tangshan Univ.* **2017**, *30*, 21–24.
24. Qi, X.F.; ShangGuan, S.T.; Zhang, G.B. Site selection and developmental prospect of a hot dry rock resource project in the Matouying Uplift, Hebei Province. *Earth Sci. Front.* **2020**, *27*, 9.
25. Lei, H.; Xu, T. TOUGH2-BIOT: A coupled thermal-hydrodynamic-mechanical model for geothermal development. *Trans.-Geotherm. Resour. Counc.* **2014**, *38*, 305–312.
26. Lei, H.; Xu, T.; Jin, G. TOUGH2-BIOT–A simulator for coupled thermal–hydrodynamic–mechanical processes in subsurface flow systems: Application to CO2 geological storage and geothermal development. *Comput. Geosci.* **2015**, *77*, 8–19. [CrossRef]
27. Pruess, K.; Oldenburg, C.M.; Moridis, G.J. *TOUGH2 User's Guide Version 2*; Office of Scientific & Technical Information Technical Reports; Office of Scientific & Technical Information: Oak Ridge, TN, USA, 1999.
28. Yuan, Y.; Xu, T.; Moore, J.; Lei, H.; Feng, B. Coupled Thermo–Hydro–Mechanical Modeling of Hydro-Shearing Stimulation in an Enhanced Geothermal System in the Raft River Geothermal Field, USA. *Rock Mech. Rock Eng.* **2020**, 53. [CrossRef]

29. Zhou, Z.F. *Theory on Dynamics of Fuids in Fractured Medium;* Higher Education Press: Beijing, China, 2007.
30. Xie, L.; Min, K.B. Initiation and propagation of fracture shearing during hydraulic stimulation in enhanced geothermal system. *Geothermics* **2016**, *59*, 107–120. [CrossRef]
31. Mcclure, M.W.; Horne, R.N. Investigation of injection-induced seismicity using a coupled fluid flow and rate/state friction model. *Geophysics* **2012**, *76*, 34–35. [CrossRef]
32. Dempsey, D.; Kelkar, S.; Lewis, K.; Hickman, S.; Davatzes, N.; Moos, D.; Zemach, E. Modeling Shear Stimulation of the Desert Peak EGS Well 27-15 Using a Coupled Thermal-Hydrological-Mechanical Simulator. In Proceedings of the 47th US Rock Mechanics/Geomechanics Symposium, San Francisco, CA, USA, 23–26 June 2013.
33. McGarr, A. On relating apparent stress to the stress causing earthquake fault slip. *J. Geophys. Res. Solid Earth* **1999**, *104*, 3003–3011. [CrossRef]
34. Lee, H.S.; Cho, T.F. Hydraulic Characteristics of Rough Fractures in Linear Flow under Normal and Shear Load. *Rock Mech. Rock Eng.* **2002**, *35*, 299–318. [CrossRef]

energies

MDPI

Essay

Comparative Analysis of Heating Efficiency of a Single-Well Geothermal System in the Cold Region of Northeast China

Bo Feng [1], Hao Ren [1], Yabin Yang [2,3,*], Zhenpeng Cui [1] and Jichu Zhao [2,3]

[1] College of New Energy and Environment, Jilin University, Changchun 130021, China
[2] The Second Institute of Hydrogeology and Engineering Geology, Shandong Provincial Bureau of Geology & Mineral Resources(Lubei Geo-Engineering Exploration Institute), Dezhou 253072, China
[3] Shandong Provincial Research Center of Geothermal Resources and Reinjection, Dezhou 253072, China
* Correspondence: echoyabin@163.com

Abstract: Geothermal energy is a type of renewable energy that has rich reserves, is clean, environmentally friendly and has been widely used in the heating industry. The single-well closed-loop geothermal system is a technology with the characteristics of "taking heat without taking water" and is mainly used for geothermal energy heating. Although the heating requirements in the cold region of Northeast China are urgent, the traditional heating mode not only has high economic costs but also causes serious damage to the environment. Therefore, it is of important practical significance to change the heating structure and develop and utilize geothermal energy for heating according to local conditions. In this study, the actual operating single-well geothermal system in the Songyuan area of Jilin Province is used as a case study, and a numerical model is established based on the T2WELL simulation program. The flow production temperature and heat extraction response law of the single-well system in the M1 and M2 wells are contrasted and analyzed under the three key factors of geothermal gradient and injection temperature and flow rate. Based on the simulation results, an optimized development and utilization plan for the M1 and M2 wells is proposed. These results provide a theoretical reference and heating potential evaluation for the promotion of single-well geothermal systems in Northeast China. Taking the geothermal gradient of 4.2 ° C/hm as an example, after 30 years of operation, the heat extraction of the M1 well is 406 kW, and that of the M2 well is 589 kW. Compared with the M1 well, although the M2 well has higher heat extraction, the radial variation in reservoir temperature is more than 50 m under long-term operation, which is not conducive to long-term development and utilization.

Keywords: geothermal energy heating; single-well geothermal systems; T2WELL; geothermal gradient; optimized development

1. Introduction

With climate change becoming a potentially irreversible threat to society, renewable energy has attracted worldwide attention. Geothermal energy is a clean, sustainable and widely-distributed form of renewable energy, which has been developed and utilized on a large scale by more than 80 countries in the form of heat pumps, space heating, bathing and power generation [1]. The development and utilization of geothermal resources can effectively reduce greenhouse gas emissions, which is important for achieving the transformation of energy structure and the "double carbon" strategic goal [2–4]. Northeast China has long and cold winters, and therefore, an urgent need for heating, but traditional heating methods create environmental problems, while also bearing high heating costs [5,6]. Heating methods urgently need to be moved in a clean, low-carbon and economical direction [7]. On 30 November 2022, the Energy Administration of Jilin Province released the "14th Five-Year Plan for Controlling Total Coal Consumption in Jilin Province". This plan states that geothermal energy resources should be developed scientifically and utilized in an orderly manner according to local conditions. In industrial development zones, tourist attractions, new residential areas, government-invested public buildings and other areas, the use of medium- and deep-level geothermal energy will be carried out to create a "geothermal Sanxia of the whole region". This shows that the promotion of geothermal energy heating in the cold areas of Northeast China is of great significance.

At present, geothermal energy development includes shallow ground source heat pump, medium and deep buried pipe and deep artificial thermal storage fracturing technologies [8]. The medium and deep buried pipe technology includes single-well, U-shaped-well and doublet-well systems [9,10] (Figure 1). Single-well closed-loop geothermal systems have the feature of "taking heat but not water", and the heat exchange of coaxial casing can be realized by circulating the injected fluid in the closed system [11]. Compared with the ground source heat pump technology, the heat exchange fluid of single-well systems does not enter the ground but only flows between the wellbore and the original groundwater flow; chemical and stress fields are not damaged. Compared with U-shaped-well and doublet-well systems, not only are single-well systems adaptable, but they also possess the advantages of low construction costs [12]. In addition, the operation of single-well systems is not dependent on climatic conditions, which ensures the long-term, stable and efficient operation of geothermal systems [13].

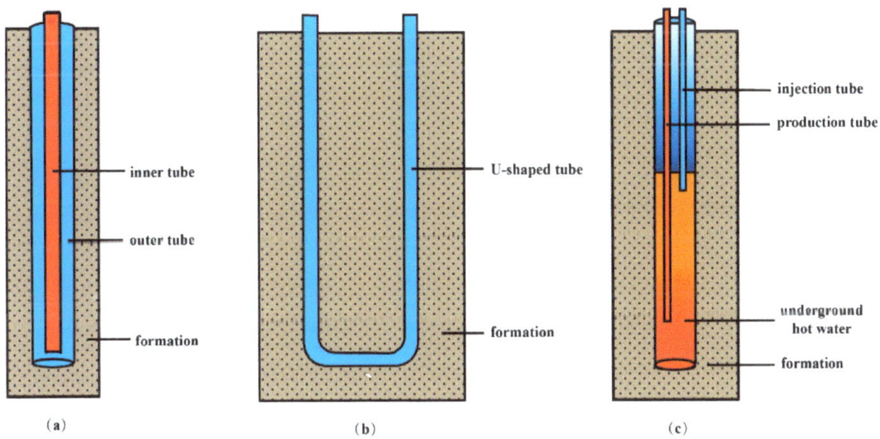

Figure 1. Schematic diagram of medium–deep buried pipe system. (**a**) Single-well system. (**b**) U-shaped well system. (**c**) Doublet-well system.

Many researchers in China and internationally have studied single-well closed-loop systems. Falcone and others [14] analyzed the advantages and disadvantages of the current development of single-well closed-loop systems. Cui and others [15] proposed a single-well enhanced geothermal system based on a single-well system in combination with hydraulic fracturing and analyzed the economic feasibility of this technology. Yu and others [16] established a new enhanced deep well heat transfer system through a concrete material with high thermal conductivity to improve the heat extraction efficiency of this technology. Bu and others [17] comprehensively analyzed the effects of insulation material properties, injection water temperature and flow rate on the system performance by establishing the flow heat transfer equation of the fluid in the geothermal well and the energy equation of the rock. Song and others [18] studied and analyzed the coaxial casing closed-cycle heat extraction technology in Xiongan New Area by combining numerical simulations and field trials and made an economic analysis of the different thermal storage conditions and insulation structures based on the field conditions. Hu and others [19] studied the fluid flow and thermal processes of CO_2 instead of water in the single-well closed-loop systems, and the heat-extracting mechanism was analyzed. A new Antoine–

based correlation for a water–CO_2 mixture was proposed for geothermal applications by Niknam and others [20]. Leontidis and others [21] carried out modeling of the reinjection of two-phase non-condensable gases and water in geothermal wells. All of the above discussions have provided an important reference and ideas for the establishment of models in this research.

For the cold area of Northeast China, it is of great practical significance to accelerate the change in heating structure and develop and utilize geothermal energy for heating in accordance with local conditions. However, the development of single-well geothermal systems in the cold region of Northeast China is still lacking sufficient theoretical guidance. Therefore, in this study, a conceptual model was established by a numerical simulation method, and the existing single-well system in the Songyuan area was used as a reference. The T2WELL simulation program was used to analyze the heating efficiency of the single-well closed-loop geothermal system in terms of geothermal gradient, water injection temperature and flow rate and other elements. The response of each element was optimized by combining the simulation results, so as to propose a reasonable development plan for a single-well closed-loop geothermal system in a well group, which provided a theoretical basis and a reference direction for future research on the large-scale development of a single-well closed-loop geothermal system in the cold region of Northeast China.

2. Study Area

In this study, the existing single-well system in the Songyuan area was used as a reference. The geological conditions of the Songyuan area are thin in the Cenozoic Era, where the total thickness of the Quaternary and Neogene systems is only 100 m. The lower strata are, in order, the Upper Cretaceous Nenjiang, Yaojia, Qingshankou, Quantou and Dengluku groups. From the surface to a depth of 2100 m, the main body of the formation is mudstone with a small amount of sandstone. Below 2100 m, the main body of the lithology is granite basement (Figure 2) [22].

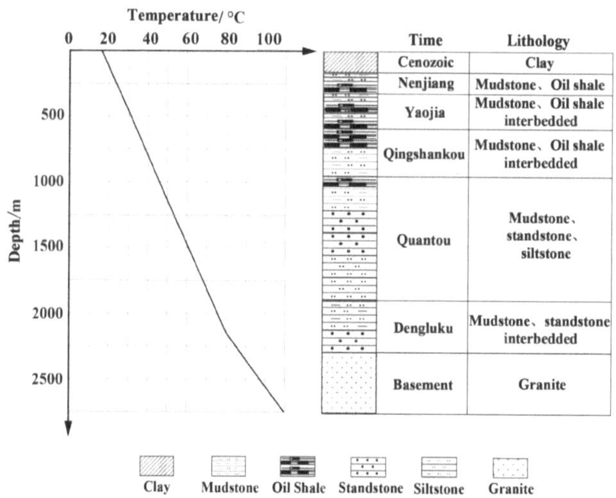

Figure 2. Lithology and well temperature records of research domain [22].

From the data of previous studies and logging data, the temperature measurements in the Songliao Basin at a depth of 3000 m below the ground are in the range of 101–138 °C with an average temperature of 124 °C [23]. The average geothermal gradient is 3.8 °C/hm. The average heat flow of Songliao Basin is ~73–79 mW/m^2, significantly higher than the national average of 61.5 mW/m^2 [24]. It can be seen that the Songyuan area is rich in geothermal resources and has excellent potential for heating with the potential for the large-scale development of geothermal energy.

In addition to its abundant geothermal resources, the Songyuan area has extremely rich experience in the development and utilization of single-well closed-loop geothermal systems. At present, the single-well system located in the Songyuan Economic Development Zone, Jilin Province, has been operating successfully for 4 years, with a cumulative heating area of 10,000 m^2 within the site and economic benefits of RMB 5.3 million, thereby providing valuable experience for the large-scale promotion of geothermal single wells in the Songyuan area.

3. Fundamentals and Mathematical Model

3.1. Fundamentals

The basic principle of a single-well closed-loop geothermal system is shown in Figure 3. The system consists of three parts: production wells, injection wells and a rock reservoir [25]. The heat exchange fluid exchanges heat with the formation through the injection well, and the hot water is pumped from the bottom of the well to the ground through the insulated production well for use. The annular hollow part in the middle of the injection and the production well is used as the water injection port. The production well is used as the extraction channel, and the two wells are connected at the bottom. The fluid flows in the injection well and is pumped out from the production well after reaching the bottom of the well. The cold water is injected through the injection well for heat exchange with the reservoir. In addition, the whole underground system is completely closed without destroying the original underground flow and chemical fields.

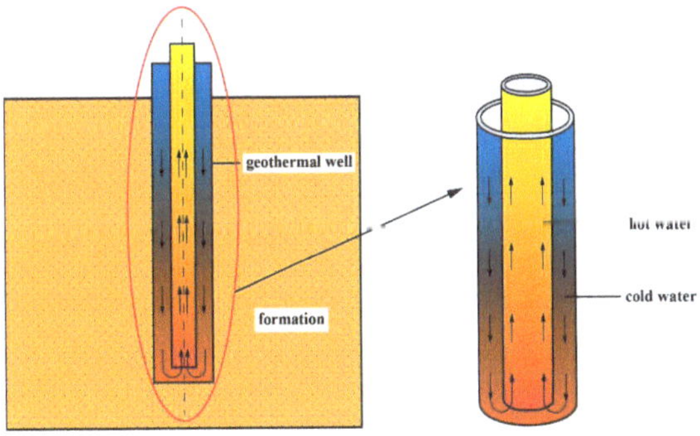

Figure 3. Schematic diagram of single-well closed-loop geothermal system.

3.2. Mathematical Model

The wellbore–reservoir coupling calculation process is the key for achieving accurate water–heat coupling numerical simulations in a single-well system. During the whole operation process, the hydrothermal migration processes involved are as follows: (1) the fluid flow and heat transfer process in injection wells; (2) the hydrothermal migration process of fluid in production wells; (3) heat exchange between fluid and reservoir during fluid migration.

During the operation of a geothermal well, the heat transfer process proceeds as follows: geothermal reservoir—cement layer wrapped outside the wellbore—injection well pipe—fluid inside the well. Therefore, the mathematical model should mainly consider the heat transfer equation between each heat transfer medium. For the analysis of the heat transfer situation, the model is assumed to be a homogeneous formation, and the energy loss during the transfer process is neglected so that the heat transferred from the rock reservoir to the wellbore is equal to the heat transferred from the wellbore to the fluid.

3.3. Fluid Flow and Heat Transfer Equation in Injection Wells

The heat exchange process in the injection well mainly occurs between the reservoir and the fluid in the injection well.

$$\frac{\partial T_R}{\partial t} + \frac{\partial (v T_R)}{\partial z} = S_{rR} + S_{RS} \tag{1}$$

$$S_{RS} = \frac{h_R 2\pi r_3 (T_{s,wall} - T_R)}{\rho A_R C_p} \tag{2}$$

where T_R is the temperature (K) of the fluid in the injection well, t is time (s), v is the fluid flow rate (m/s) in the well, z is the vertical depth (m), S_{rR} is the heat transfer (K/s) between the recovery and injection wells, S_{RS} is the heat transfer (K/s) between the fluid and the well wall, h_R is the convective heat transfer coefficient (W/(m^2·K)) of the inner well wall, r_3 is the inner radius (m) of the inner casing, $T_{s,wall}$ is the temperature (K) of the well wall in contact with the fluid, ρ is the density of water, A_R is the injection well circulation area (m^2), and C_P is the specific heat capacity (J/(kg·K)) of water.

3.4. Fluid Flow and Heat Transfer Equation in Production Wells

The heat exchange process in the production well mainly occurs in the fluid of injection well and production well.

$$\frac{\partial T_r}{\partial t} + \frac{\partial (v T_r)}{\partial z} = -S_{rR} \tag{3}$$

$$S_{rR} = \frac{k_1 (T_r - T_R)}{\rho A_r C_p} \tag{4}$$

where T_r is the temperature (K) of the fluid in the production well, k_1 is the heat transfer per unit length(W/(m·K)), and A_R is the production well circulation area (m^2).

3.5. Convective Heat Transfer Coefficient

The Convective heat transfer coefficient mainly depends on several parameters of the fluid itself.

$$\begin{aligned} h_{r2} &= 0.023\lambda \frac{Re^{0.8} Pr^{0.4}}{d_e} \\ h_{r1} &= 0.023\lambda \frac{Re^{0.8} Pr^{0.3}}{2r_1} \end{aligned} \tag{5}$$

where h_{r2} and h_{r1} are the convective heat transfer coefficients (W/(m^2·K)) of the outer and inner walls of the extraction well, respectively, λ is the thermal conductivity (W/(m·K)) of the convective liquid, Re is the Reynolds number of the fluid, Pr is the Prandtl number of the fluid, d_e is the hydraulic diameter (m), and r_1 is the outer radius (m) of the inner casing.

3.6. Boundary and Initial Conditions and Initial Condition

The heat transfer from the rock to the well wall is equal to the heat transfer from the well wall to the fluid, and the contact between the three is given by the third type of boundary condition:

$$h_R (T_{s,wall} - T_R)|r = r_3 = \lambda_W \frac{\partial T_W}{\partial r}|r = r_4 \tag{6}$$

$$T_{W,0} = T_{sur} + T_g z \tag{7}$$

where λ_w is the thermal conductivity (W/(m·K)) of the rock, T_W is the rock temperature (K), r_4 is the outer radius (m) of the inner casing, $r_4 = r_3 + b_s$, b_s is the thickness (m) of the inner casing, $T_{w,0}$ is the initial temperature (K) of the rock, T_{sur} is the surface temperature (K), T_g is the geothermal gradient (K/m), and z is the distance from the surface (well depth) (m).

Since this simulation only involves the heat transfer process between the wellbore and the reservoir, the initial conditions of the simulation only consider the temperature and pressure conditions, where the temperature conditions are calculated according to the geothermal gradient selected by the scheme, and the pressure conditions are uniformly distributed according to the hydrostatic pressure.

4. Numerical Simulation

4.1. Numerical Simulation Code

The TOUGH2-WELL (T2WELL) simulation program is a tool mainly used to simulate non-isothermal multiphase flows in coupled wellbore–reservoir systems [26]. The program was developed by adding a wellbore grid to a reservoir grid based on the original software TOUGH2, thus enabling the simultaneous calculation of flow coupling in the wellbore and reservoir [27,28]. The TOUGH procedure uses the integral finite difference method and the unconditionally convergent implicit difference method for spatial discretization and temporal discretization, respectively. One of the modules, EOS1, was specifically designed for hydraulic geothermal modeling and has now been incorporated into T2WELL. In addition, a version of parallel computing that can handle the significant computational burden caused by the large number of grid meshes was developed based on the parallel program introduced by Feng and others [29].

4.2. Simulation Parameter Selection

A single-well system at a depth of 2300 m was put in place in the study area. To improve the simulation accuracy, it was necessary to use the measured data obtained from the operation of the system during the heating season. It was also essential to calibrate the lithological parameters related to the permeability, thermal conductivity and specific heat capacity of the stratum in which it was located by fitting the results obtained from the model calculations to the measured data for the flow production temperature profile under heating conditions. By debugging the relevant parameters several times, the accuracy of the model could be improved so that it could better reflect the actual engineering.

The simulation injection temperature and circulation flow rate were selected according to the actual site heating data in the Songyuan area. The main part of the study area from the surface to 2100 m underground is a mudstone layer mixed with a small amount of sandstone. The part below 2100 m is granite base, so the geological reservoir can be mainly generalized into two layers: mudstone and granite. The parameters of the two layers are different, and the initial parameters of the model are determined by referring to the relevant information and the values taken by previous authors [13,22,30], with the debugging carried out on this basis. Figure 4 shows the fit of the measured production flow temperature profile to the simulated data profile.

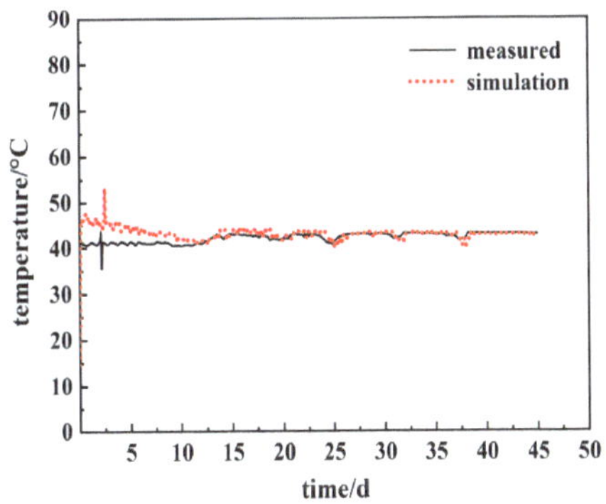

Figure 4. Produced flow temperature fitting curve.

It can be found from the fitting curve that the simulated temperature is slightly high in the early stage of model operation. This is mainly because the initial stage of the simulation is in adaptive operation, the injection flow is relatively large and the reservoir temperature is high, so the simulated temperature is high. In summary, the selected parameters of this simulation are reasonable, and the fit is good. The final selected simulation parameters are shown in Table 1.

Table 1. Model numerical simulation parameters.

Media	Lithology	Major Parameters	Parameter Values
Formation parameters (Pore media)	Mudstone	Porosity	0.25
		Heat conductivity W/(m·K)	3.2
		Specific heat capacity J/(kg·K)	950
		Horizontal permeability (mD)	1.5
	Granite	Porosity	0.05
		Heat conductivity W/(m·K)	3.5
		Specific heat capacity J/(kg·K)	970
		Horizontal permeability (mD)	1.6
Wellbore parameters		Diameter mm	178
		Porosity	0.01
		Heat conductivity W/(m·K)	1.5

4.3. Simulation Strategy

The main performance indicators of a single-well geothermal system include the outlet water temperature, heat extraction (thermal power) and thermal reservoir temperature [31]. The main factors affecting these three indicators are well depth, well diameter, geothermal gradient, reservoir thermal conductivity, run time, injection temperature and mass flow rate. This study combines the actual engineering demand and realistic heating needs in the Songyuan area, mainly for the simulation under two well depth conditions of 2500 and 3000 m. The depth of the M1 well is 2500 m, and the depth of the M2 well is 3000 m, and the heat exchange law analysis is carried out by combining three key conditions of geothermal gradient, water injection temperature and injection flow rate. By summarizing the above simulations, we finally propose a reasonable development plan for the single-well closed-loop geothermal system in the cold region of Northeast China. Table 2 presents the set up for the specific simulation scenario.

Table 2. Numerical simulation strategy.

Well	Depth (m)	Geothermal Gradient (°C/hm)	Injection Temperature (°C)	Injection Flow (m^3/h)	Operating Time (a)
M1	2500	2.5, 3, 3.5, 4, 4.2	20, 25, 30	20, 30, 40	30
M2	3000				

5. Analysis of Factors Influencing Production Temperature and Heat Extraction

5.1. Influence of Geothermal Gradient

According to the injection temperature of 25 °C and injection flow rate of 30 m^3/h, the rest of the simulation parameters were selected according to the previous selection. The simulation results curves of the flow production temperature and heat extraction for different geothermal gradients for the M1 and M2 wells are given in Figure 5.

As can be seen from Figure 5, the M1 well produced temperatures of 31.5, 33.2, 34.9, 36.6 and 37.4 °C in the first year of operation at geothermal gradients of 2.5, 3.0, 3.5, 4.0 and 4.2 °C/hm with heat extractions of 225, 285, 345, 405 and 431 kW, respectively. The production temperatures for continuous operation up to the 30th year were 30.3, 31.6, 32.9, 34.2 and 34.8 °C. The corresponding heat extractions were only 184, 230, 275, 321 and 341 kW, with extraction decreases of 18.2%, 19.3%, 20.3%, 20.7% and 20.9%, respectively.

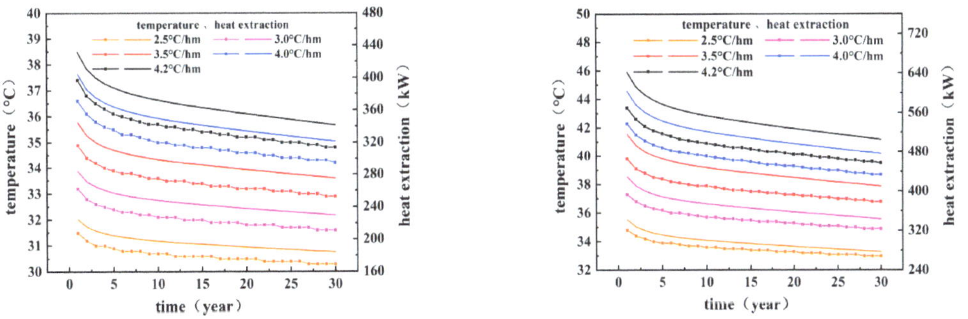

(a) M1 well

(b) M2 well

Figure 5. Comparison of production temperature and heat extraction under different geothermal gradients.

It can be seen that in the early stage of geothermal well operation, the production flow temperature is high, and a substantial heat extraction can be obtained, but as the operation time continues, the production flow temperature gradually decreases, and the heat extraction also decreases. The main reason for this is that as the operation time continues, the heat of reservoir is continuously absorbed, and the reservoir temperature continues to decrease, so the production flow temperature and heat extraction also decrease. The lower geothermal gradient has relatively good stability under long-term operation, and the extraction drop is relatively small, but the flow-producing temperature and heat extraction are lower due to the lower reservoir temperature. The reason for this is that the reservoir temperature is low, and the temperature drop is relatively small. As the geothermal gradient increases, the heat extraction of geothermal wells rises very significantly, so places with large geothermal gradients are more suitable for single-well geothermal development.

The M2 well could reach 342, 429, 516, 603 and 642 kW of heat extraction in the first year of operation at the above five geothermal gradients. The corresponding heat extractions for continuous operation up to the 30th year were 277, 344, 410, 476 and 505 kW with heat extraction decreases of 19.0%, 19.8%, 20.5%, 21.1% and 21.3%, respectively. The increase in heat extraction was significant when the well depth increased. Compared to the M1 well, the heat extraction gains in the M2 well were 117, 144, 171, 198 and 211 kW at five geothermal gradients in the first year of operation and 93, 114, 135, 155 and 164 kW after 30 years of operation, respectively. By comparing the M1 and M2 wells, it can be found that the increase in well depth increased the flow path of the heat transfer fluid in the wellbore, resulting in a significant increase in the heat transfer area and therefore a significant increase in the produced flow temperature and heat extraction. However, the decrease in heat extraction under long-term operation was larger compared to the lower well depth, which may not be conducive to sustainability.

After this comparison, it can be found that an increase in well depth can make up for a low geothermal gradient, and the development of single-well geothermal systems in places with low geothermal gradients should appropriately increase the depth of the geothermal wells. Simultaneously, the higher the geothermal gradient, the greater the power gain obtained from the increase in well depth, but the long-term operational stability is relatively poor, and the heat extraction drop is large.

5.2. *Influence of Injection Temperature*

According to the geothermal gradient of 4.0 °C/hm and the injection flow rate of 30 m^3/h, the rest of the simulation parameters were selected according to the previous section. The simulation curves of the produced flow temperature and heat extraction for different injection temperatures for the M1 and M2 wells are shown in Figure 6.

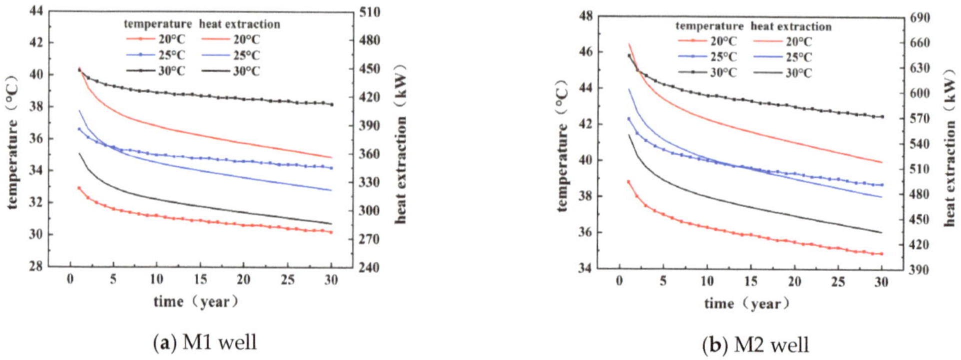

Figure 6. Comparison of production temperature and heat extraction at different injection temperatures.

The M1 well produced water at 32.9, 36.6 and 40.3 °C and 450, 405 and 360 kW of heat in the first year of operation at injection temperatures of 20, 25 and 30 °C, respectively. After 30 years of continuous operation, the production temperatures were 30.2, 34.2 and 38.2 °C, and the heat extractions were 356, 321 and 286 kW with extraction decreases of 20.9%, 20.7% and 20.6%, respectively.

In the first year of operation, the M2 well produced water at 38.8, 42.3 and 45.8 °C and 657, 603 and 549 kW of heat at injection temperatures of 20, 25 and 30 °C, respectively. After 30 years of continuous operation, the production temperatures were 34.9, 38.7 and 42.5 °C, and the heat extractions were 518, 476 and 434 kW with extraction decreases of 21.2%, 21.1% and 20.9%, respectively.

These results show that as the injection temperature increased, the production flow temperature increased in both the M1 and M2 wells, showing a positive correlation, but the heat extraction then decreased, showing a negative correlation. The higher the temperature of the injected water, the more difficult it is to exchange heat with the reservoir, and it may not be able to fully absorb the heat, which is the main reason for this phenomenon. Higher injection temperatures not only play an important role in maintaining the stability of geothermal wells for long-term continuous operation (multiyear operating extraction reduction) but also allow for higher effluent temperatures, albeit at the expense of heat extraction. The lower the injection temperature, the higher the heat extraction and the more favorable for heating, but this does not mean that a single lower injection temperature is optimal. A low injection temperature also leads to a limited increase in the temperature of the water coming out, meaning that the process of extracting heat after the water is pumped out of the geothermal well becomes difficult, and the heat extraction process may be more costly.

5.3. Influence of Injection Flow

According to the geothermal gradient of 4.0 °C/hm and the injection temperature of 25 °C, the rest of the simulation parameters are selected according to the previous section. The simulation curves of the flow production temperature and heat extraction at different injection temperatures for the M1 and M2 wells are shown in Figure 7.

In the first year of operation, the production temperatures of the M1 well were 40.7, 36.6 and 34.4 °C, and the heat extractions were 369, 405 and 439 kW at injection flow rates of 20, 30 and 40 m^3/h, respectively. After 30 years of continuous operation, the production temperatures were 37.4, 34.2 and 32.5 °C, and the heat extractions were 291, 321 and 352 kW with extraction decreases of 21.1%, 20.7% and 19.8%, respectively.

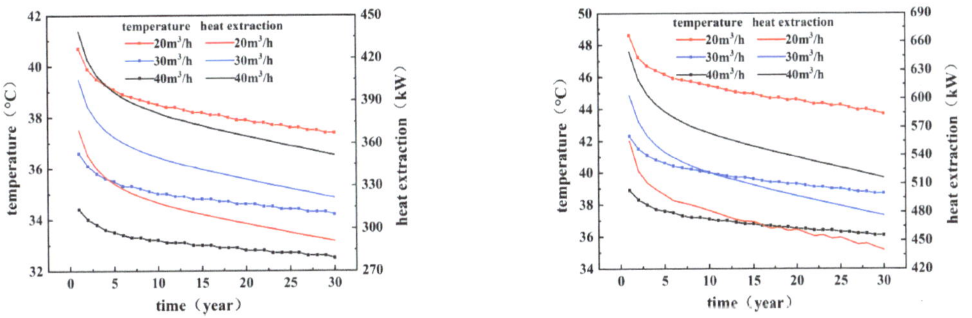

(**a**) M1 well (**b**) M2 well

Figure 7. Comparison of production temperature and heat extraction at different injection flows.

In the first year of operation, the production temperatures of the M2 well were 48.6, 42.3 and 38.9 °C, and the heat extractions were 556, 603 and 649 kW at injection flow rates of 20, 30 and 40 m^3/h, respectively. After 30 years of continuous operation, the output water temperatures were 43.7, 38.7 and 36.1 °C, and the heat production powers were 435, 476 and 516 kW with power decreases of 21.8%, 21.1% and 20.5%, respectively.

In summary, it can be found that the injection flow rate is negatively correlated with the temperature of the produced flow, and the higher the flow rate, the lower the temperature of the produced flow. The higher the injection flow rate, the more water that is exchanged with the reservoir for heat exchange per unit time, which also leads to a lower temperature of produced flow. However, there is a positive correlation with heat extraction, and the change in heat extraction is very obvious as the flow rate gradually increases. The high injection flow rate can improve the stability under continuous operation for many years, and the extraction variation under multiyear operation is relatively small. Although a larger injection flow rate will produce higher heat extraction, in actual engineering, the diameter of a single well is usually small, basically within 1 m, which cannot guarantee an injection flow rate that is too large. An excessive flow rate will also accelerate the decay of the flow production temperature with time, which is not conducive to the sustainable development and utilization of geothermal single wells.

5.4. Geothermal Energy Optimization

Summarizing the above multiple sets of simulation scenarios, patterns can be derived from three key factors:

(1) The geothermal gradient is positively correlated with the flow production temperature and heat extraction of the geothermal wells. The development of a single-well system is more effective in areas with high geothermal gradients, while areas with low geothermal gradients require deeper wells to increase their development value.

(2) The injection temperature is positively correlated with the flow production temperature of the geothermal well and negatively correlated with the heat extraction. Too high an injection temperature is not conducive for obtaining substantial heat extraction, while too low a temperature increases the difficulty of thermal extraction.

(3) The injected flow rate is positively correlated with the heat extraction of the single-well system and negatively correlated with the produced flow temperature. The higher the flow rate, the higher the heat extraction, but the injected flow rate should not be too high, otherwise it will make the temperature of the produced flow lower and the thermal extraction difficult.

Among the three influencing factors, the geothermal gradient has the greatest influence, followed by the injection temperature and the injection flow rate with the least influence. The geothermal gradient is limited by geothermal geological conditions and cannot be changed artificially. Thus, if we want to increase the heating area of a single geothermal well, we need to lower the injection temperature and increase the injection flow rate as much as possible. The above three laws were integrated, assuming that 100,000 m^2 of heating was required; a single-well closed-loop geothermal system was selected for heating; and two kinds of wells, M1 and M2, are used as the starting points, combined with key factors, so as to propose the optimal layout of the single-well system for heating, in order to achieve the optimal working conditions.

5.5. Analysis of M1 Well Heating Potential

In this simulation, an injection temperature of 20 °C and an injection flow rate of 40 m^3/h were chosen. The simulation results under five geothermal gradients are shown in Figure 8.

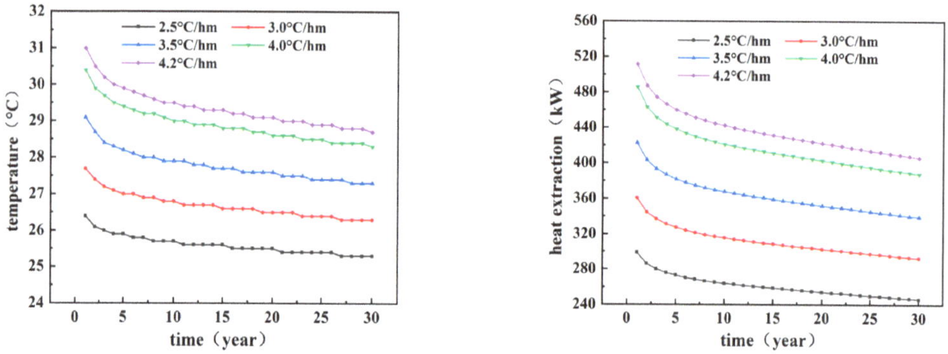

Figure 8. Optimized production temperature and heat extraction for M1 well at different geothermal gradients.

From Figure 8, it can be found that when the geothermal gradient is 2.5 °C/hm, the heat extraction is 299 kW in the first year of operation and 246 kW after 30 years of operation. According to the national heating standard of 38 W/m^2, the heating area can be 7868 m^2 in the first year and 6474 m^2 after 30 years of continuous operation. When the geothermal gradient is 4.2 °C/hm, the heat extraction is 511 kW in the first year of operation and 406 kW after 30 years of operation, and the heating area is 13,447 m^2 in the first year and 10,684 W/m^2 after 30 years of continuous operation. The specific simulation results are shown in Table 3.

Table 3. Simulation results of M1 well under different geothermal gradients.

Geothermal Gradient °C/hm	Production Temperature °C (1st/30th)	Heat Extraction kW (1st/30th)	Heating Area m^2 (1st/30th)	Extraction Decay %
2.5	26.4 25.3	299 246	7868 6474	17.8
3.0	27.7 26.3	360 293	9474 7711	18.6
3.5	29.1 27.3	423 339	11,132 8921	19.8
4.0	30.4 28.3	486 388	12,789 10,211	20.2
4.2	31.0 28.7	511 406	13,447 10,684	20.6

From these simulation results, it can be found that the M1 well is relatively small in depth and cannot obtain a high flow production temperature and heat extraction in areas with a low geothermal gradient, which makes the development of the M1 well unsuitable. It is more valuable to develop the well in areas with a high geothermal gradient that is 1.7 times more the heating area at 4.2 °C/hm than at 2.5 °C/hm. Therefore, the M1 well is more suitable for areas with high geothermal gradients.

5.6. Analysis of M2 Well Heating Potential

The same injection temperature and flow rate as for the M1 well were used in this simulation. The simulation results under five geothermal gradients are shown in Figure 9.

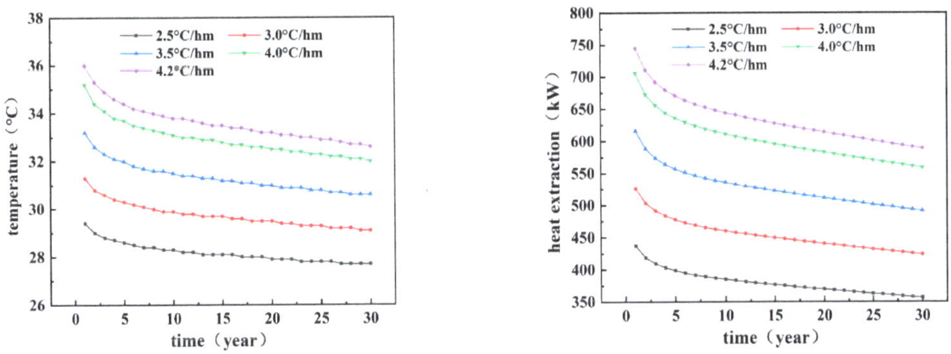

Figure 9. Optimized production temperature and heat extraction for M2 well at different geothermal gradients.

We found that when the geothermal gradient is 2.5 °C/hm, the heat extraction is 437 kW in the first year of operation and 357 kW after 30 years of operation. According to the national heating standard of 38 W/m^2, the heating area can be 7868 m^2 in the first year and 6474 m^2 after 30 years of continuous operation. When the geothermal gradient is 4.2 °C/hm, the heat extraction is 511 kW in the first year of operation and 406 kW after 30 years of operation, and the heating area is 13,447 m^2 in the first year and 10,684 m^2 after 30 years of continuous operation. The specific simulation results are shown in Table 4.

Table 4. Simulation results of M2 well under different geothermal gradients.

Geothermal Gradient °C/hm	Production Temperature °C (1st/30th)	Heat Extraction kW (1st/30th)	Heating Area m^2 (1st/30th)	Extraction Decay %
2.5	29.4 27.7	437 357	11,500 9394	18.3
3.0	31.3 29.1	526 424	13,842 11,158	19.4
3.5	33.2 30.6	616 492	16,211 12,947	20.1
4.0 9474	35.2 32.0	706 559	18,579 14,711	20.8
4.2	36.0 32.6	745 589	19,605 15,500	20.9

The relatively large depth of the M2 well allows for a higher flow production temperature and heat extraction under the same conditions compared to the M1 well, but the increased depth of the well also makes the well more expensive to drill.

5.7. Design of Optimal Development Plan

A comprehensive analysis of the above two wells led to the following patterns:

(1) M1 wells are more suitable for development in areas with high geothermal gradients.

(2) M2 wells are more suitable for development in areas with low geothermal gradients.

(3) Lowering the injection temperature and increasing the injection flow rate as much as possible can obtain more heat extraction for heating.

If at least 100,000 m^2 of heating is required, the M1 or M2 well is selected for development for different geothermal gradients, and a simple static payback period estimate is made on this basis [32]. Table 5 shows the design of the development schemes under different geothermal gradients.

Table 5. Design of development schemes under different geothermal gradients.

	Geothermal Gradient (°C/hm)	Number of Wells	Heating Area 30-Year Average (m^2)	Well Completion Cost (Million Yuan)	Heating Charge (Million Yuan/Heating Season)	Payback Period (Year)
M1	2.5	15	103,018	160	330	7.3
	3.0	13	106,725		342	6.1
	3.5	11	105,043		336	5.2
	4.0	10	109,369		350	4.6
	4.2	9	103,279		330	4.4
M2	2.5	11	109,973	210	352	6.6
	3.0	9	107,534		344	5.5
	3.5	8	111,194		356	4.7
	4.0	7	110,947		355	4.1
	4.2	6	100,215		321	3.9

It could be found that although the cost of drilling a single geothermal well is lower for the M1 well compared to the M2 well, the number of wells required to meet the heating demand of at least 100,000 m^2 needs to be increased accordingly, so the total cost would be relatively higher, and the payback period would be relatively longer.

Although the payback period of the M2 well is short, the service life of geothermal wells may also be shortened due to long-term high-flow operation, so the change in geothermal field temperature before and after the extraction of geothermal wells is analyzed using a geothermal gradient of 4.2 °C/hm as an example (Figure 10).

It could be found that after 30 years of continuous mining in the M1 well, the temperature of each temperature zone decreases more obviously and the radius of influence gradually increases with depth, but it can still barely maintain the original temperature of each temperature zone, and the development can still be continued under such mining intensity. However, after the M2 well was subjected to long-term continuous mining, the temperature change in the reservoir was very obvious, and the radius of influence was significant. The reservoir temperature in the near-well area decreased by more than one third, and it became very difficult to develop again under the original mining intensity. This is also the reason why heat extraction produces a significant decrease. This also reflects the fact that although the M2 well can obtain more heat extraction, it comes at the cost of reduced service life.

The results of the above analysis show that the M1 well is more suitable for areas with high geothermal gradients, and the M2 well is more suitable for areas with low geothermal gradients. Areas with high geothermal gradients can make up for the lack of well depth by their own higher reservoir temperature, while areas with low geothermal gradients can fill the lack of their own reservoir temperature by increasing the well depth. If the heating demand of 100,000 m^2 is to be met, the input cost of the M1 well is relatively large and the payback period is long, but the temperature change in the reservoir is relatively small, and the service life of the geothermal well is longer, which is more suitable for sustainable development. The input cost of the M2 well is smaller, and the payback period is shorter, but the temperature change in the reservoir is significant, and it is relatively difficult to meet the needs of long-term development and utilization.

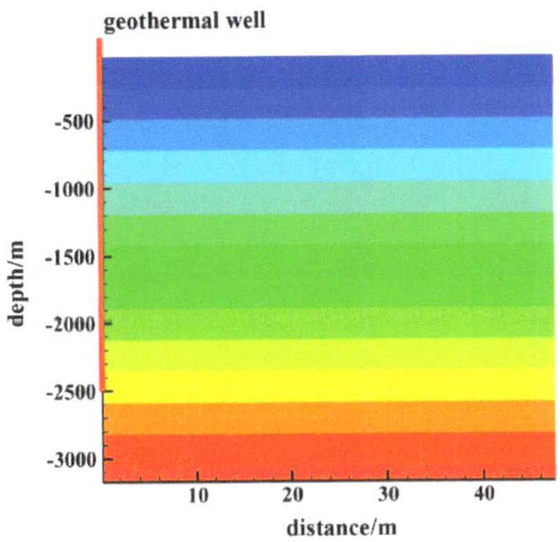

(**a**) Before operating (M1)

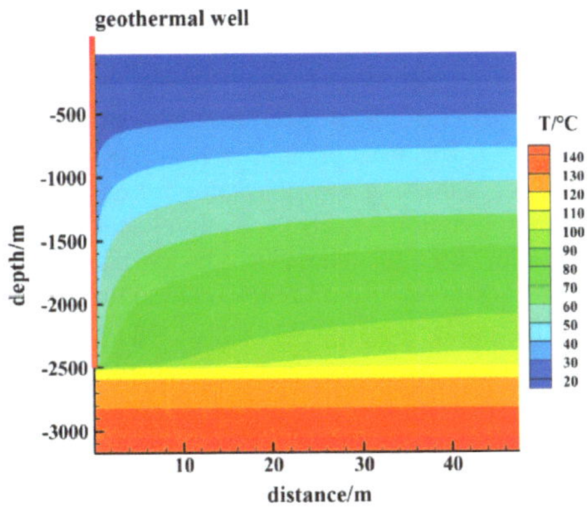

(**b**) After 30 years of operation (M1)

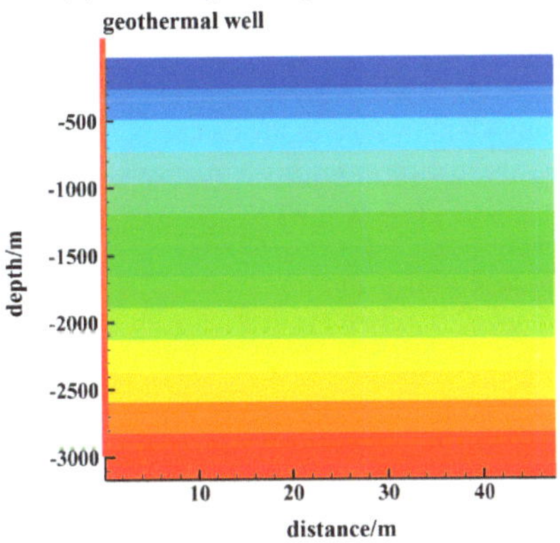

(**c**) Before operating (M2)

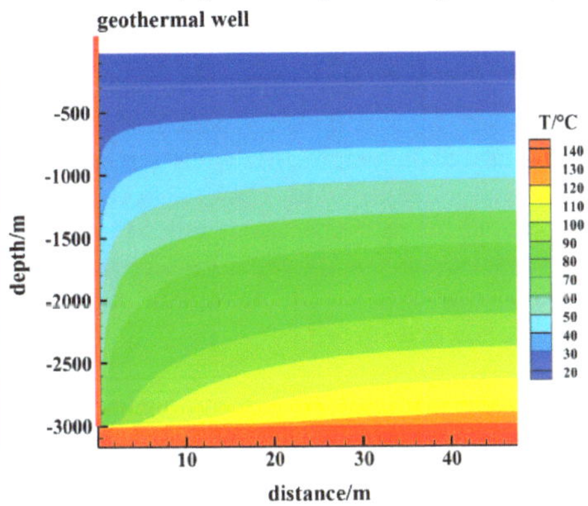

(**d**) After 30 years of operation (M2)

Figure 10. Temperature field diagram of M1 and M2 wells before and after operation at 4.2 °C/hm geothermal gradient.

In addition, reasonable well spacing is also an important factor affecting the sustainable development of the single-well geothermal system. It can be inferred from Figure 10 that the influence radius of temperature change under long-term continuous exploitation is large for both M1 and M2 wells. If the well spacing is too small (less than 50 m), the heat efficiency of multiple groups of single wells will be significantly reduced. Therefore, when developing multiple groups of single-well systems, the well spacing should be as large as possible. For example, the temperature influence radius of the M1 well is smaller than that of the M2 well, so the well spacing of the M1 well can be appropriately smaller than that of the M2 well, but it should not be too small. According to previous studies [13], under such a scale of exploitation, the well spacing should be at least not less than 60 m. If conditions permit, the well spacing can be increased to more than 100 m. This measure is also more conducive to the long-term development and utilization of the single-well system.

6. Conclusions

To summarize the research in this study, the following conclusions can be obtained:

(1) The development of single-well systems is very dependent on geological conditions. When the ground temperature gradient is 2.5 °C/hm, the heat extraction of the M1 well is about 225 kW, and that of the M2 well is about 342 kW. When the ground temperature gradient reaches 4.2 °C/hm, it can reach 431 kW and 642 kW, respectively. The geothermal gradient is positively correlated with the flow production temperature and heat extraction of single-well systems, and the larger the geothermal gradient, the higher the corresponding flow production temperature and heat extraction.

(2) The injection temperature is positively correlated with the production flow temperature of the single-well system but negatively correlated with the heat extraction, while the injection flow rate is negatively correlated with the production flow temperature and positively correlated with the heat extraction. The effect of injection temperature on the heat extraction is more pronounced than that of the injection flow rate. Although lowering the injection temperature and increasing the injection flow rate can increase the heat extraction, it will lower the production flow temperature, and the low production flow temperature will increase the difficulty of heat extraction.

(3) Areas with high geothermal gradients are more suitable for the M1 well with smaller well depths due to higher reservoir temperatures, while extending the service life of single-well systems. However, if a single well-system is to be developed in areas with low geothermal gradients, a larger-depth M2 well is more appropriate due to the lower reservoir temperature. For example, when the geothermal gradient is higher than 3.5 °C/hm, it may be more appropriate to select the M1 well with a long service life.

(4) To meet the heating demand of 100,000 m^2, the cost of developing M1 wells is relatively high, and the payback period is long, but a longer service life can be obtained. The cost of developing M2 wells is relatively low, and the payback investment period is short, but the useful life may be short. Taking the 4.2 °C/hm geothermal gradient as an example, the payback period of the M1 well is 4.4 years, and that of the M2 well is 3.9 years. However, the radial temperature response of the reservoir is more obvious. Therefore, the well spacing should be guaranteed to be at least 60 m when large-scale well cluster deployment is carried out.

Author Contributions: Conceptual model and numerical simulation, B.F. and H.R.; Visualization and writing, Z.C.; Funding acquisition and resource, J.Z.; Project administration, Y.Y. All authors have read and agreed to the published version of the manuscript.

Funding: This work was supported by the National Natural Science Foundation of China (No: 42072331, U1906209).

Data Availability Statement: The original contributions presented in the study are included in the article. Further inquiries can be directed to the corresponding author.

Conflicts of Interest: The authors declare no conflict of interest.

References

1. Lund, J.W.; Toth, A.N. Direct utilization of geothermal energy 2020 worldwide review. *Geothermics* **2021**, *90*, 101915. [CrossRef]
2. Daarnhouwer, M. Assessing the Potential of Deplete Gasfields for Geothermal Energy. Ph.D. Thesis, Delft University of Technology, Delft, The Netherlands, 2013.
3. Bruhn, D.; Jolie, E.; Huenges, E. European research efforts on engineered and superhot geothermal systems within horizon2020 (Conference Paper). *Trans.-Geotherm. Resour. Counc.* **2018**, *42*, 2381–2395.
4. Tian, J.; Pang, Z.; Li, Y.; Zhou, X. Research progress on geothermal gas. *Acta Geol. Sin.* **2022**, *96*, 1752–1766. (In Chinese)
5. Wang, G.; Zhang, W.; Liang, J.; Lin, W.; Liu, Z.; Wang, W. Evaluation of Geothermal Resources Potential in China. *Acta Geophys. Sin.* **2017**, *38*, 449–450+134+451–459. (In Chinese)
6. Neves, R.; Cho, H.; Zhang, J. Techno-economic analysis of geothermal system in residential building in Memphis, Tennessee. *J. Build. Eng.* **2020**, *27*, 100993. [CrossRef]
7. Wang, S.; Kang, R.; Feng, X.; Wang, K.; Fang, C.; Cao, Q.; Cui, Z. A optimization method of heat pump peak-shaving ratio based on economic evaluation of geothermal heating project. *Nat. Gas Ind.* **2021**, *41*, 152–159. (In Chinese)
8. Noorollahi, Y.; Saeidi, R.; Mohammadi, M.; Amiri, A.; Hosseinzadeh, M. The effects of ground heat exchanger parameters changes on geothermal heat pump performance—A review. *Appl. Therm. Eng.* **2018**, *129*, 1645–1658. [CrossRef]

9. Ye, H.; Li, Q.; Sun, F. Numerical simulation of CO_2 circulating in a retrofitted geothermal well. *J. Pet. Sci. Eng.* **2019**, *172*, 217–227. [CrossRef]
10. Sun, F.; Yao, Y.; Li, G.; Li, X. Geothermal energy development by circulating CO_2 in a U-shaped closed loop geothermal system. *Energy Convers. Manag.* **2018**, *174*, 971–982. [CrossRef]
11. Bu, X.; Jiang, K. Performance of a geothermal single well for continuous and intermittent heating. *Chin. Sci. Technol. Sci.* **2019**, *49*, 1514–1522. (In Chinese)
12. Xu, T.; Hu, Z.; Feng, B.; Feng, G.; Li, F.; Jiang, Z. Numerical evaluation of building heating potential from a co-axial closed-loop geothermal system using wellbore-reservoir coupling numerical model. *Energy Explor. Exploit.* **2020**, *38*, 733–754. [CrossRef]
13. Feng, B.; Liu, X.; Zhang, G.; ShangGuan, S.; Hu, Z.; Yuan, Y.; Feng, G. Numerical simulation on the sustainable development potential of a single-well closed-cycle geothermal system. *Nat. Gas Ind.* **2020**, *40*, 146–155. (In Chinese)
14. Falcone, G.; Liu, X.; Okech, R.R.; Seyidov, F.; Teodoriu, C. Assessment of deep geothermal energy exploitation methods: The need for novel single-well solutions. *Energy* **2018**, *160*, 54–63. [CrossRef]
15. Cui, G.; Wang, W.; Dou, B.; Liu, Y.; Tian, H.; Zheng, J.; Liu, Y. Geothermal Energy Exploitation and Power Generation via a Single Vertical Well Combined with Hydraulic Fracturing. *J. Energy Eng.* **2022**, *148*, 04021058. [CrossRef]
16. Yu, H.; Xu, T.; Yuan, Y.; Gherardi, F.; Feng, B.; Jiang, Z.; Hu, Z. Enhanced heat extraction for deep borehole heat exchanger through the jet grouting method using high thermal conductivity material. *Renew. Energy* **2021**, *177*, 1102–1115. [CrossRef]
17. Bu, X.; Ran, Y.; Li, H.; Lei, J.; Wang, L. Enhancing heat transfer methods of existing geothermal single well heating system. *J. Sol. Energy* **2020**, *41*, 369–374. (In Chinese)
18. Song, X.; Zhang, Y.; Li, G.; Li, R.; Yu, C.; Li, J.; Guo, X. Performance Study of the Downhole Coaxial Closed-Loop Heat Exchange Technology in Xiong'an New Area. *J. Tianjin Univ. (Sci. Technol.)* **2021**, *54*, 971–981. (In Chinese)
19. Hu, Z.X.; Xu, T.F.; Feng, B.; Yuan, Y.L.; Li, F.Y.; Feng, G.H.; Jiang, Z.J. Thermal and fluid processes in a closed-loop geothermal system using CO_2 as a working fluid. *Renew. Energy* **2020**, *154*, 351–367. [CrossRef]
20. Niknam, P.H.; Talluri, L.; Fiaschi, D.; Manfrida, G. Sensitivity analysis and dynamic modelling of the reinjection process in a binary cycle geothermal power plant of Larderello area. *Energy* **2021**, *214*, 118869. [CrossRef]
21. Leontidis, V.; Niknam, P.H.; Durgut, I.; Talluri, L.; Manfrida, G.; Fiaschi, D.; Akin, S.; Gainville, M. Modelling reinjection of two-phase non-condensable gases and water in geothermal wells. *Appl. Therm. Eng.* **2023**, *223*, 120018. [CrossRef]
22. Xie, Y. Research on Medium-Deep Geothermal Heating Potential and Model Assessment in Songyuan, Jilin. Ph.D. Thesis, Jilin University, Changchun, China, 2019.
23. Hu, S.; He, L.; Wang, J. *Compilation of Heat Flow Data in the China Continental Area*, 3rd ed.; Division of Science and Technology and Achievement Transformation, Institute of Geology and Geophysics, Chinese Academy of Sciences: Beijing, China, 2001; Volume 16.
24. Jiang, G.; Gao, P.; Rao, S.; Zhang, L.; Tang, X.; Huang, F.; Zhao, P.; Pang, Z.; He, L.; Hu, S.; et al. Compilation of heat flow data in the continental area of China(4th edition). *Chin. J. Geophys.* **2016**, *59*, 2892–2910. (In Chinese)
25. Bu, X.; Ran, Y.; Wang, L.; Lei, J.; Li, H. Analysis of key factors affecting single well geothermal heating. *J. Zhejiang Univ. (Eng. Sci.)* **2019**, *53*, 957–964. (In Chinese)
26. Pan, L.; Oldenburg, C.M.; Wu, Y.S.; Pruess, K. *T2Well/ECO2N Version 1.0: Multiphase and Non-Isothermal Model for Coupled Wellbore-Reservoir Flow of Carbon Dioxide and Variable Salinity Water*; University of California: Berkeley, CA, USA, 2011.
27. Li, M.; Gou, Y.; Hou, Z.; Were, P. Investigation of a new HDR system with horizontal wells and multiple fractures using the coupled wellbore-reservoir simulator TOUGH2MP-WELL/EOS3. *Environ. Earth Sci.* **2015**, *73*, 6047–6058. [CrossRef]
28. Xu, T.; Yuan, Y.; Jia, X.; Lei, Y.; Li, S.; Feng, B.; Hou, Z.; Jiang, Z. Prospects of power generation from an enhanced geothermal system by water circulation through two horizontal wells: A case study in the Gonghe Basin, Qinghai Province, China. *Energy* **2018**, *148*, 196–207. [CrossRef]
29. Feng, G.; Xu, T.; Gherardi, F.; Jiang, Z.; Bellani, S. Geothermal assessment of the Pisa plain, Italy: Coupled thermal and hydraulic modeling. *Renew. Energy* **2017**, *111*, 416–427. [CrossRef]
30. Peng, W. Characteristics of Thermal Reservoir and Evaluation of Geothermal Resources in the First Member of Quantou Formation in Songyuan Area. Master's Thesis, Jilin University, Jilin, China, 2022.
31. Kong, Y.; Chen, C.; Shao, H.; Pang, Z.; Xiong, L.; Wang, J. Principle and capacity quantification of deep-borehole heat exchangers. *Chin. J. Geophys.* **2017**, *60*, 4741–4752. (In Chinese)
32. Chen, Z.; Osadetz, K.G.; Chen, X. Economic appraisal of shale gas resources, an example from the Horn River shale gas play, Canada. *Pet. Sci.* **2015**, *12*, 712–725. [CrossRef]

energies

MDPI

Article

Effects of Water–Rock Interaction on the Permeability of the Near-Well Reservoir in an Enhanced Geothermal System

Bo Feng [1], Zhenpeng Cui [1], Xiyao Liu [1], Shuantong Shangguan [2], Xiaofei Qi [2] and Shengtao Li [3,*]

[1] College of Environment and Resources, Jilin University, Changchun 130021, China
[2] No. 2 Exploration Team Hebei Bureau of Coal Geological Exploration, Xingtai 054000, China
[3] Center for Hydrogeology and Environmental Geology Survey, China Geological Survey, Baoding 071051, China
* Correspondence: li.st@163.com

Abstract: During the operation of an enhanced geothermal system (EGS), the non-equilibrium temperature, pressure, and hydrochemistry caused by fluid injection intensify water–rock interactions, induce the mineral dissolution and precipitation in the reservoir near an injection well (also referred to as the near-well reservoir), and change reservoir permeability, thus affecting continuous and efficient geothermal exploitation. Based on the investigation of the M-1 injection well of the EGS in the Matouying uplift of Hebei Province, China, a THC reactive solute transport model using the TOUGHREACT program was established in this study to explore the mineral dissolution and precipitation laws of the near-well reservoir and their influencing mechanisms on the reservoir porosity and permeability in the long-term fluid injection of this well. As indicated by the results, the dissolution of primary feldspar and chlorite and the precipitation of secondary minerals (mainly dolomite and illite) occurred and water–rock interaction significantly reduced the porosity and permeability of the near-well reservoir in the long-term continuous injection process. Appropriate reduction in the injection flow rate, injection temperature, and the Mg^{2+} and K^{+} contents in the injected water can help inhibit the formation of secondary minerals and delay the plugging process of the near-well reservoir.

Keywords: enhanced geothermal system; water–rock interaction; secondary precipitation; reactive solute transport; TOUGHREACT

1. Introduction

The high dependence on fossil fuels in today's society not only leads to a global energy crisis but also further aggravates challenges such as greenhouse gas emissions and environmental pollution [1,2]. The use of new clean energy will help solve these problems [3,4]. Compared with traditional energy sources and other renewable energy sources, geothermal energy has many engineering and environmental advantages, such as non-seasonal dependence and a small area for resource extraction, and is thus an ideal new clean energy [5,6]. It can be used for many purposes, such as heating and power generation [7]. At the current energy consumption rate, it is estimated that the global energy demand in 2800 can be met as long as 1% of the geothermal resources in the crust are fully exploited and utilized [8,9].

An EGS refers to an artificial geothermal system in which geothermal energy is economically extracted from low-permeability rock masses by artificially creating geothermal reservoirs [10]. During geothermal exploitation using an EGS, circulating fluids (usually water) are injected through injection wells to make them move along the fracture network and exchange heat with surrounding rocks, and then high-temperature fluids can be extracted from production wells for power generation and comprehensive utilization [11]. This process involves sufficient contact between the low-temperature injected fluids and the high-temperature rock masses in deep reservoirs. However, relevant studies have shown that the changes in temperature, pressure, and hydrochemistry caused by the heat transfer fluids lead to a sustained non-equilibrium between solids and solutes and further drive fluid–rock interactions, leading to different degrees of mineral dissolution and precipitation which change the permeability of the near-well reservoir and affect sustainable and efficient geothermal exploitation [12].

Because of the difficulty, long period, and high costs of actual operation and monitoring, numerical simulation technology is considered an effective means to conduct site-scale analysis and prediction. In recent years, researchers in related fields have carried out massive laboratory experiments and numerical simulations to explore the dissolution and precipitation of minerals induced by the interactions between circulating fluids and geothermal reservoir rocks and the resultant changes in geothermal reservoir permeability during the production and operation of an open-loop geothermal system. Borgia et al. analyzed the feasibility of using CO_2 as the transfer fluids of an EGS through numerical simulations, and the results showed that the injection of CO_2 mixed with water inhibited secondary precipitation and effectively prolonged the operating life of the EGS but limited the heat extraction rate [12]. Driba et al. established a one-dimensional THC reactive solute transport model based on Phreeqc and OpenGeosys to explore the effects of the chemical composition of injected fluids on the deep geothermal reservoir permeability during the reinjection of low-temperature brine [13]. Ren et al. proposed a mixed discretization method based on embedded meshes and accurately and effectively determined the coupling process of multiphase flow and geomechanics in fractured reservoirs [14]. Salimzadeh et al. established a 3D multifield coupling model for fractured geothermal reservoirs and explored the influence of low-temperature fluid injection on fracture channel deformation [15]. Pei et al. established a new flow and geomechanical coupling model and effectively captured the fracture deformation process in unconventional fractured reservoirs [16]. Li et al. improved the calculation efficiency while maintaining relatively high accuracy by combining the embedded discrete fracture model (EDFM) and the extended finite element method (XFEM) [17]. These studies provide references and bases for further improving and expanding the reactive solute transport model [18].

Regenspurg et al. carried out a hydrogeochemistry equilibrium calculation based on the collection, testing, and analysis of numerous samples to explore the cause and mechanism of the severe plugging of production wells at the site in Groß Schönebeck, Germany, revealing that frequent well shutdown constantly made the metal-rich fluids oversaturated in the process of reacting with the wellbore and cooling, thus exacerbating the plugging of geothermal wells [19]. Based on a large number of experimental studies, as well as reinjection experiments and hydrogeochemical simulations, Ma et al. proved that chemical plugging caused by carbonate and silicate was the main reason for the significant decrease in the injection flow rate of Well Xianyang 2 over time [20]. Yanaze et al. established a geochemical plugging model for the Sumikawa Geothermal Power Plant to predict the decrease in permeability caused by the deposition of silica scale on the site and found that reducing the silica concentration and the injection flow rate can effectively delay the decrease in the reservoir permeability [21]. Zhao et al. carried out high-temperature and high-pressure static experiments and hydrogeochemical simulations based on the PHREEQC program to determine the interactions between different injected water bodies and high-temperature granites, finding that using low-salinity water as the injected water could reduce the possibility of geothermal reservoir plugging [22]. Ke et al. used similar methods to evaluate the effect of different fluids on the release of typical minerals in the FORGE site [23].

The secondary precipitation caused by fluid injection has been analyzed in the above studies. However, most of these studies employed small-scale static simulations and experiments, which cannot reflect the dynamic process occurring in actual reservoirs. Moreover, there is still a lack of targeted research on the effects of dynamic near-well water–rock interactions and solute transport on reservoir permeability during long-term injection. Based on the investigation of injection well M-1 in the EGS in the Matouying uplift of Hebei Province, China, this study established a THC reactive solute transport model using the TOUGHREACT program and analyzed and predicted the laws of the mineral dissolution and precipitation caused by water-rock reactions and solute transport and their influence on the near-well reservoir permeability in the long-term injection of this well. This study provides a technical reference and theoretical basis for the long-term stable operation of geothermal systems.

2. Geological Conditions of Regional Geothermal Resources

The study area is located in the Matouying uplift of Leting County, Tangshan City, eastern Hebei Province [24]. The Matouying uplift is a significant positive secondary tectonic unit within the Huanghua depression in North China. It is bounded by the Mabei fault and connected to the Leting sag in the north, bounded by the Bogezhuang fault and connected to the Nanpu sag in the west, and is bounded by the Hongfangfang fault and adjacent to the Shijiutuo sag in the south (Figure 1). The tectonic pattern consisting of the high uplift and deep sags creates favorable geological conditions for the upward migration of deep heat flow and the convergence of lateral heat flow toward the uplift area [25]. The Matouying uplift has terrestrial heat flow of greater than 75 mW/m^2, which is higher than the global average, making this uplift a favorable prospect area for the exploitation of deep geothermal resources in Hebei Province [26].

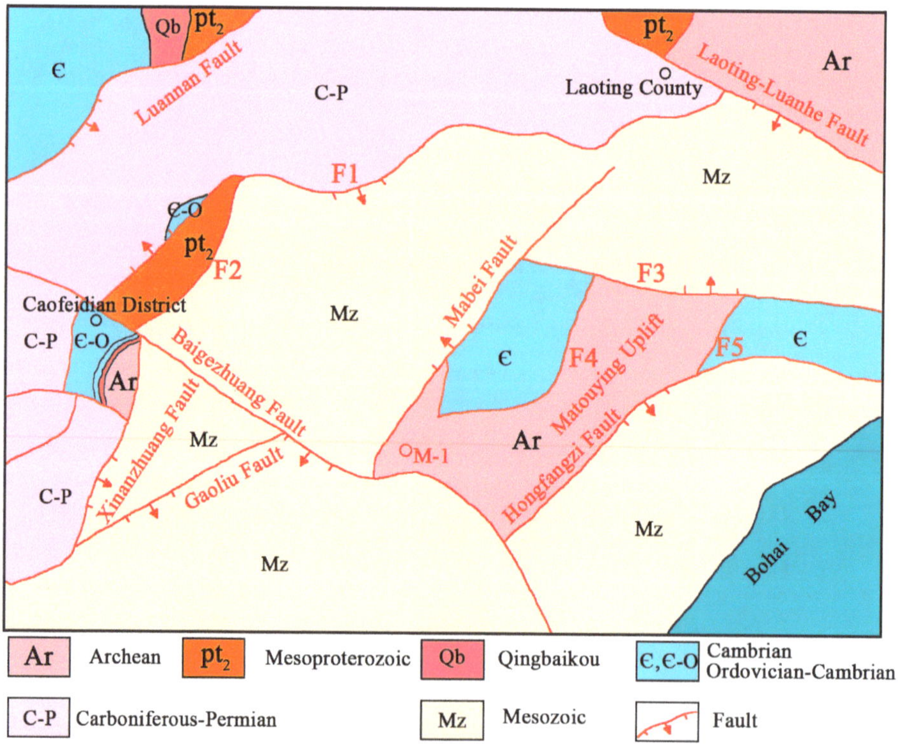

Figure 1. Tectonic location of the study area.

The EGS project in the Matouying uplift is located in southern Dazhuanghe Village, Caofeidian District, Tangshan City. This project consists of injection well M-1 and production well M-2, which are about 200 m apart. Well M-1 has a final hole depth of 4502.11 m, with the water filtering interval at a depth of 4188.08–4502.11 m. This well has a water temperature of up to 150 °C at a depth of 3960 m. The main strata encountered during the drilling of this well include the Quaternary (Q), the Neogene Minghuazhen (N_2m) and Guantao formations (N_1g), and the Archean Baimiao Formation (A*rb*). The Cenozoic strata in the upper part of this well mainly include clay, sandstones, and mudstones, while the Archean Baimiao Formation in the lower part is mainly composed of metamorphic granodiorites, metamorphic tonalites, metamorphic hornblende monzogranites, and biotite plagiogneisses, indicating apparent migmatization [27].

3. Numerical Simulation Method

3.1. Simulator

The simulator used in this study was the TOUGHREACT program developed by thc Lawrence Berkeley National Laboratory. By introducing the reactive solute transport process into the basic framework of the simulator TOUGH2 for non-isothermal flows of multicomponent and multiphase fluids, this simulator couples the flow, heat transfer, solute transport, and geochemical reaction processes of underground multiphase fluids while considering the thermophysicochemical processes under conditions of different temperatures, pressure, water saturation, and ionic strength [28]. TOUGHREACT is suitable for one-, two-, and three-dimensional porous and fractured media with physical and chemical heterogeneity, as well as any number of gas-, liquid-, and solid-phase chemicals. Therefore, this program is widely applied to the exploitation and utilization of geothermal energy, the geological storage of carbon dioxide, pollutant migration and remediation, and nuclear waste disposal [29].

3.2. Governing Equations

The main governing equations used by TOUGHREACT are introduced as follows. For numerical simulations, the mass and energy conservation equation is the most basic and core equation, and its generalized expression is shown in Equation (1).

$$\frac{d}{dt}\int_{V_n} M^K dV = \int_{\Gamma_n} F^K \cdot n d\Gamma + \int_{V_n} q^K dV \tag{1}$$

where $t[s]$ is the time; n is the current grid; κ is the different components; Γ_n [m^2] is the grid connection area; V_n [m^3] is the grid volume; M^K, F^K, and q^K are the cumulative item, migration (i.e., inflow or outflow) item, and source-sink item of mass or energy, respectively.

Chemical transport equations (derived from mass conservation) have the same structure as fluid and heat flow equations, and the equation of multicomponent chemical transport in the liquid phase is expressed as Equation (2).

$$\begin{aligned} &\frac{\Delta t}{V_n}\sum_m A_{nm}\left[u_{nm}^{k+1} C_{nm}^{(j),\ k=1,s+\frac{1}{2}} + D_{nm}\frac{C_m^{(j),k=1,s+\frac{1}{2}} - C_n^{(j+1),k=1,s+\frac{1}{2}}}{d_{nm}}\right] \\ &= \Delta M_n^{(j),k+1} - q_n^{(j),k+1}\Delta t - R_n^{(j),k+1,s}\Delta t \qquad J = 1,2,\ldots,N_c \end{aligned} \tag{2}$$

The transport of aqueous and gaseous species through advection and molecular diffusion are considered for both liquid and gas phases in this simulator. Acid-base and redox are considered under the local equilibrium assumption. Depending on the computer memory and CPU performance, any number of chemical species in the liquid, gas, and solid phases can be accommodated. The dissolution and precipitation processes of minerals are characterized by two mechanisms, i.e., thermodynamic equilibrium and reaction kinetic equilibrium. The thermodynamic governing equations used by TOUREACT are shown in Equations (3) and (4), and the governing equations of reaction kinetics based on the transition state theory are shown in Equations (5) and (6).

$$SI_m = \log_{10}\Omega_m = 0 \tag{3}$$

$$\Omega_m = K_m^{-1} \prod_{j=1}^{Nc} c_j^{v_{mj}} \gamma_j^{v_{mj}} \quad m = 1, \ldots Np \tag{4}$$

$$r = A_m k_m \left[1 - \left(\frac{Q}{K}\right)^{\mu}\right]^n \tag{5}$$

$$k = k_{25}^{nu} exp\left[\frac{-E_a^{nu}}{R}\left(\frac{1}{T} - \frac{1}{298.15}\right)\right] + k_{25}^{H} exp\left[\frac{-E_a^{H}}{R}\left(\frac{1}{T} - \frac{1}{298.15}\right)\right] \alpha_H^{n_H} + k_{25}^{OH} exp\left[\frac{-E_a^{OH}}{R}\left(\frac{1}{T} - \frac{1}{298.15}\right)\right] \alpha_{OH}^{n_{OH}} \tag{6}$$

where m is the mineral serial number; SI is the saturation index of minerals; Ω is the saturation of minerals; K is the thermodynamic equilibrium constant; γ is the activity coefficient; r [mol/s] is the mineral reaction rate (it is positive for dissolution and negative for precipitation); A [g/cm^2] is the reaction specific surface area of minerals; k [mol/(L·s)] is the reaction rate constant related to temperature; Q is the reaction quotient; μ and n are laboratory empirical parameters; k_{25} [mol/(L·s)] is the reaction rate constant at 25 °C; Ea [KJ/mol] is the activation energy; R [J/mol/K] is the gas constant; T [K] is the absolute temperature, and α [mol/L] is the activity.

The dissolution and precipitation of minerals lead to changes in porosity and permeability, which further affect the fluid flow and solute transport process. Therefore, it is necessary to accurately describe the dynamic changes in porosity and permeability. When the numerical simulations of water-rock reactions are performed using TOUGHREACT, the medium porosity is determined using Equation (7).

$$\varphi = 1 - \sum_{m=1}^{nm} fr_m - fr_u \tag{7}$$

where φ is porosity; nm is the number of mineral species; fr_m is the volume fraction of mineral m; and fr_u is the volume fraction of unreactive rock.

As fr_m changes, the medium porosity is recalculated at each time step. TOUGHREACT provides a variety of options for calculating permeability variation caused by mineral dissolution and precipitation. Some options only depend on changes in porosity, while others are formulas related to changes in fracture pore size and pore throat diameter. In this study, the cubic law (Equation (8)) was used to describe the relationship between permeability variation and porosity:

$$K = k_i \left(\frac{\varphi}{\varphi i}\right)^3 \tag{8}$$

where k [m^2] is the permeability; and k_i [m^2] and φ_i are the initial permeability and porosity, respectively.

4. Model Establishment

4.1. Conceptual Model

This study aims to explore the laws of mineral dissolution and precipitation caused by fluid injection and their influencing mechanisms on the permeability of the near-well reservoir during the long-term operation of the EGS in the Matouying area. Multi-field coupling models of various minerals and hydrochemical components require massive calculations in the long-term simulation. Therefore, to effectively control the calculation time and improve the simulation efficiency, this study simplified the conceptual model and only simulated a 10 m thick monolayer in the water filtering interval of Well M-1. In addition, the fracture network in the artificial reservoir is mainly composed of dense small-scale fractures as revealed by the tracer test results of the Matouying EGS field (not published yet). Therefore, an equivalent porous medium was adopted in this study.

As shown in Figure 2, the model was established using the radial grid method (RZ2D) [30], with a vertical height of 10 m and a horizontal length of 200 m (i.e., the distance between the injection well and the production well on site). Considering that the influence of the injected fluids on reservoirs gradually weakens with an increase in the distance from the injection well, the grid size gradually increased in the radial direction from the center of the model. The bottom of the model was set as an impermeable boundary with a constant temperature and pressure, and the injection well was set as the Dirichlet boundary to allow the flow rate to vary with time. In the process of simulation using TOUGHREACT, the grid volume of the matrix layers was set to infinity to make the matrix temperature and pressure constant. The permeability of the matrix layers was set to 0. This setting restricted the fluid exchange process between grids but did not affect the temperature and pressure transmission. Considering that some of the continuously injected fluids return to the surface through the production well and the remaining fluids flow toward distant strata and are lost during the long-term operation of the EGS, the outer boundary of the model was set as an infinite volume boundary. Based on the research results of Qi et al. [31], the temperature and pressure conditions of the model were set to 160 °C and 40 MPa, respectively.

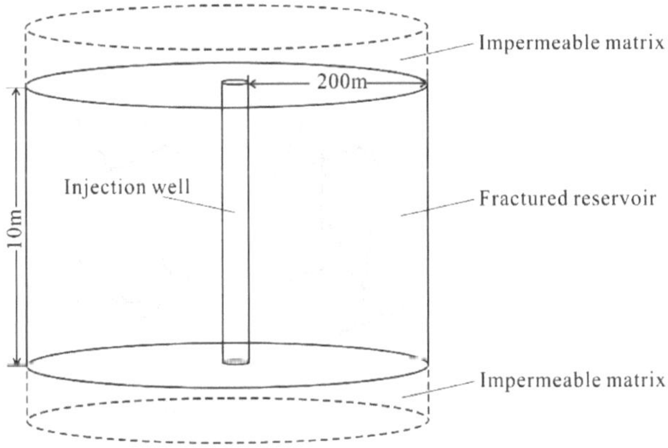

Figure 2. Conceptual model.

The chemical composition of the injected water and the rock mineral composition of the model were set according to the water-quality total-analysis data of the site and the XRD test results of rock samples, respectively [32]. The injected fluids were shallow groundwater extracted from the site and their main chemical compositions is shown in Table 1. The main minerals of the reservoir rocks included K-feldspar (30%), albites (33%), quartz (20%), chlorite (15%), and other trace minerals (2%).

Table 1. Main chemical composition of injected water.

Cation	Content (mg/L)	Proportion	Anion	Content (mg/L)	Proportion
K^+	18.20	4.17%	F^-	1.28	0.16%
Na^+	225.49	51.71%	Cl^-	295.03	36.86%
Ca^{2+}	111.96	25.68%	NO_3^-	6.91	0.86%
Mg^{2+}	31.48	7.22%	SO_4^-	169.41	21.16%
Fe^{2+}	0.54	0.12%	HCO_3^-	327.86	40.96%
Zn^{2+}	48.39	11.10%			

4.2. Parameter Setting

The density, thermal conductivity, and specific heat capacity of the geothermal reservoir rocks were set according to the research data of the Matouying area [33], and the porosity and permeability were set according to the relevant numerical simulations [34]. The petrophysical parameters used in the model are shown in Table 2. It should be noted that the porosity of the impermeable bedrock was not set to 0 but to a very low value compared to the reservoir porosity to prevent errors in the calculation process (e.g., a zero divisor). However, because the permeability was 0, this setting did not affect the flow and solute transport process in the reservoir grids.

Table 2. Petrophysical parameters.

Parameter	Value
Density (kg/m^3)	2750.00
Porosity:	
Fractured reservoir	0.50
Impermeable matrix	0.01
Permeability (m^2):	
Fractured reservoir	6.99×10^{-14}
Impermeable matrix	0
Thermal conductivity (W/kg·m)	2.20
Specific heat capacity (J/K·kg)	794.00

The dynamic reactions between minerals and the injected fluids were controlled using the parameters of mineral reaction kinetics. By referencing the numerical simulations of Na et al. [35] and Yang et al. [36], this study set the parameters of mineral reaction kinetics under different reaction mechanisms, as shown in Table 3.

Table 3. Parameters of mineral reaction kinetics.

Mineral	Neutral Mechanism		Acid Mechanism			Base Mechanism			A
	K_{25}	E	K_{25}	E	n(H+)	K_{25}	E	n(H+)	
	(mol/m^2/s)	(KJ/mol)	(mol/m^2/s)	(KJ/mol)		(mol/m^2/s)	(KJ/mol)		(cm^2/g)
Quartz	1.203×10^{-14}	87.70							9.8
K-f × 10ldspar	3.890×10^{-13}	38.00	8.710×10^{-11}	51.70	0.500	6.310×10^{-12}	94.10	−0.823	9.8
Albit × 10	2.754×10^{-13}	69.80	6.918×10^{-11}	65.00	0.457	2.512×10^{-16}	71.00	−0.572	9.8
Chlorit × 10	3.020×10^{-13}	88.00	7.762×10^{-12}	88.00	0.500				151.6
Montmor-Na	1.660×10^{-13}	35.00	1.047×10^{-11}	23.60	0.340	3.020×10^{-17}	58.90	−0.400	151.6
Montmor-Ca	1.660×10^{-13}	35.00	1.047×10^{-11}	23.60	0.340	3.020×10^{-17}	58.90	−0.400	151.6
Illit × 10	1.660×10^{-13}	35.00	1.047×10^{-11}	23.60	0.340	3.020×10^{-17}	58.90	−0.400	151.6
Calcit × 10	1.550×10^{-16}	23.50	6.018×10^{-13}	14.40	1.000				9.8
Kaolinit × 10	6.918×10^{-14}	22.20	4.898×10^{-12}	65.90	0.777	8.913×10^{-18}	17.90	−0.472	151.6
Montmor-Mg	1.660×10^{-13}	35.00	1.047×10^{-11}	23.60	0.340	3.020×10^{-17}	58.90	−0.400	151.6
Montmor-K	1.660×10^{-13}	35.00	1.047×10^{-11}	23.60	0.340	3.020×10^{-17}	58.90	−0.400	151.6
Dolomit × 10	2.951×10^{-8}	52.20	6.457×10^{-4}	36.10	0.500				9.8
Dawsonit × 10	1.260×10^{-9}	62.76	6.457×10^{-4}	36.10	0.500				9.8
Sid × 10rit × 10	1.260×10^{-9}	62.76	6.457×10^{-4}	36.10	0.500				9.8
Magn × 10sit × 10	4.508×10^{-11}	23.50	4.169×10^{-11}	14.40	1.000				9.8
H × 10matit × 10	2.512×10^{-15}	66.20	4.074×10^{-10}	66.20	1.000				9.8

4.3. Simulation Schemes

To explore the laws of the influences of key operating parameters (i.e., injection flow rate, injection temperature, and hydrochemical composition) on the mineral dissolution and precipitation and the permeability variation of the near-well reservoir, this study designed and operated multiple simulation schemes according to the actual situation of the site and the results from hydrochemical analysis and testing, as shown in Table 4 and Figure 3. The injection flow rate was set to 3 m^3/h, 4 m^3/h, 5 m^3/h, and 6 m^3/h in the simulation schemes (approximately equivalent to the injection flow rate of 90 m^3/h, 120 m^3/h, 150 m^3/h, and 180 m^3/h in the entire water filtering interval, respectively). In the injection process, the changes in wellhead injection temperature and wellbore heat loss affect the temperature of injection fluids when the fluids enter the reservoir. Therefore, the injection temperature was set to 50 °C, 70 °C, and 90 °C. It is noteworthy that the injection temperatures set in the simulations were the temperatures when the fluid flowed into the reservoir through the injection wellbore rather than the temperatures when fluids were injected from the surface.

Table 4. Simulation schemes of different injection flow rates and temperatures.

Scheme No.	Injection Flow Rate (m^3/h)	Injection Temperature (°C)
1	3	70
2	4	70
3 (Basic Scheme)	5	70
4	6	70
5	5	90
6	5	50

When designing different chemical composition schemes of injected water, the contents of three main metal ions (K^+, Ca^{2+}, and Mg^{2+}) were adjusted to replace the proportion of Na^+ in the metal cations of the basic scheme. Based on this, the hydrochemical equilibrium of the injected water was adjusted and recalculated. The specific settings of chemical components of the injected water in different schemes are shown in Figure 3.

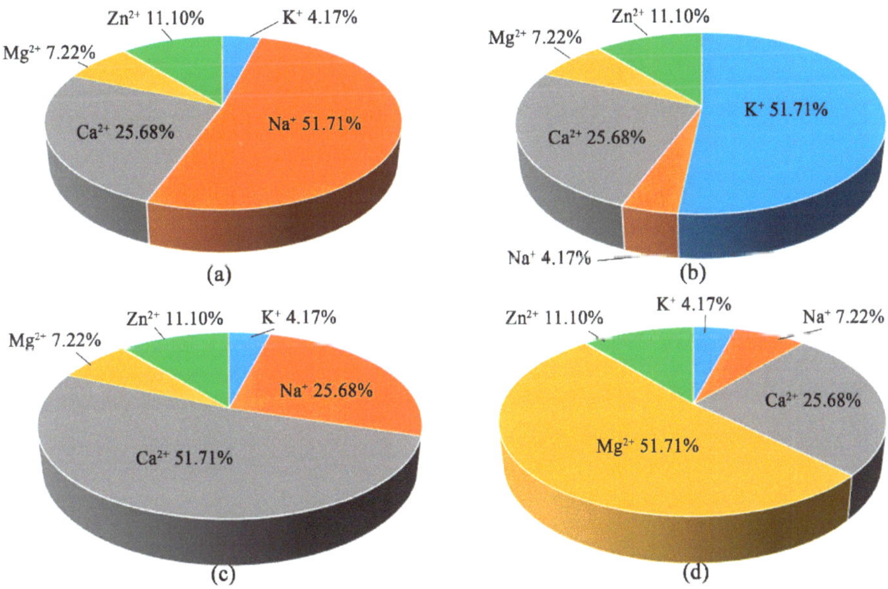

Figure 3. Simulation schemes for different chemical compositions of injected water. (**a**)—Basic scheme; (**b**)—Scheme 7; (**c**)—Scheme 8; (**d**)—Scheme 9.

5. Results and Discussion

5.1. Changes at Different Distances from the Injection Well

The changes in the reservoir porosity and permeability and initial mineral contents at different distances from the injection well after continuous injection for 10 a are shown in Figure 4. As shown in Figure 4a, after 10 a of continuous operation of the EGS, significant changes in porosity and permeability mainly occurred approximately 0–7.5 m away from the injection well. In a range within 3.5 m from the well, the reservoir porosity and permeability decreased significantly, reflecting strong precipitation of secondary minerals in this range. The decreased amplitude increased as the distance from the injection well decreased, indicating that the intensity of water-rock interactions decreased with an increase in the distance. At 10 a, the porosity at the injection well decreased by 60.0% from the initial value of 0.5 to 0.2, and the permeability decreased by 94.1% from 6.99×10^{-14} m^2 to 4.15×10^{-15} m^2, indicating that the long-term operation of geothermal wells seriously damaged the permeability of the near-well artificial reservoir. It should be noted that the reservoir permeability within the range of 3.5–7.5 from the injection point was slightly higher than its initial value, suggesting that the porosity increase caused by the initial mineral dissolution was more intense than the permeability decrease caused by the secondary mineral precipitation, as further discussed below.

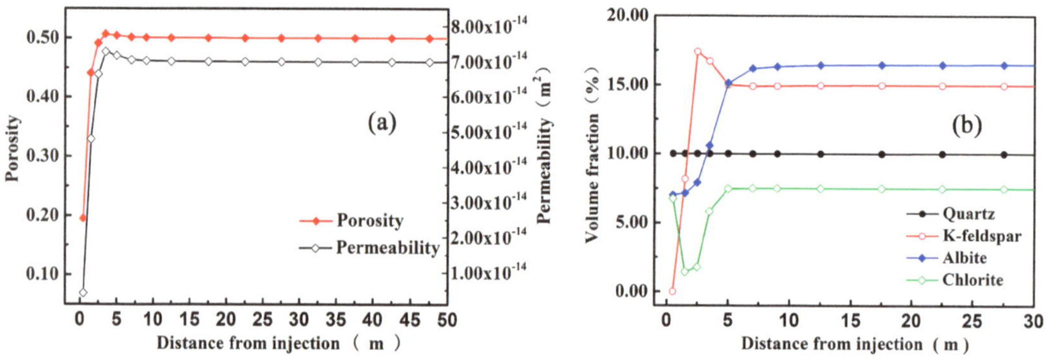

Figure 4. Simulation results at different distances from the injection point. (**a**)—Porosity and permeability; (**b**)—Content of primary minerals.

As shown in Figure 4b, the content of primary quartz in the reservoir scarcely changed during the long-term injection, while the K-feldspar, albite, and chlorite contents changed to varying degrees, reflecting that the injected fluids had a significant effect on the dissolution of the three primary minerals (Equations (9)–(11)). The albite and chlorite contents were lower than their initial values 5–7 m away from the injection well. The K-feldspar content decreased significantly 0–2.5 m away from the injection well but was higher than its initial value in the range of 2.5–7.5 m from the well, reflecting the severe dissolution of K-feldspar near the injection well and the formation of secondary K-feldspar, respectively. This changing trend is consistent with the research results of Zhao et al. [22].

$$NaAlSi_3O_8 + H_2O \rightarrow Na^+ + Al(OH)_{4^-} + H_4SiO_4 \tag{9}$$

$$KAlSi_3O_8 + H_2O \rightarrow K^+ + Al(OH)_{4^-} + H_4SiO_4 \tag{10}$$

$$Chlorite + H_2O \rightarrow 5Mg^{2+} + Al(OH)_{4^-} + H_4SiO_4 \tag{11}$$

The global secondary mineral generation at 10 a is shown in Figure 5. According to this figure, a variety of secondary minerals set in the model were generated to different degrees. Among them, illite and dolomite showed the most significant secondary precipitation, especially near the injection point (within the range of about 5 m), and the volume fraction of their precipitation reached the same order of magnitude as their initial content. Therefore, they had a far greater impact on reservoir porosity and permeability than other secondary minerals. The main reason for a large quantity of secondary dolomite is that the injected fluid contained a lot of Ca^{2+} and $HCO_3{}^-$, in which $HCO_3{}^-$ decomposed at a high temperature to produce $CO_3{}^{2-}$ or was combined with Ca^{2+} and Mg^{2+} to form dolomite precipitation (Equation (12)). In addition, the chlorite dissolution (Equation (11)) also provided more sufficient Mg^{2+}. The illite formation was mainly controlled by feldspar minerals, especially K^+ and other components provided by the K-feldspar dissolution.

$$Ca^{2+} + Mg^{2+} + CO_3{}^{2-} \rightarrow CaMg(CO_3)_2 \tag{12}$$

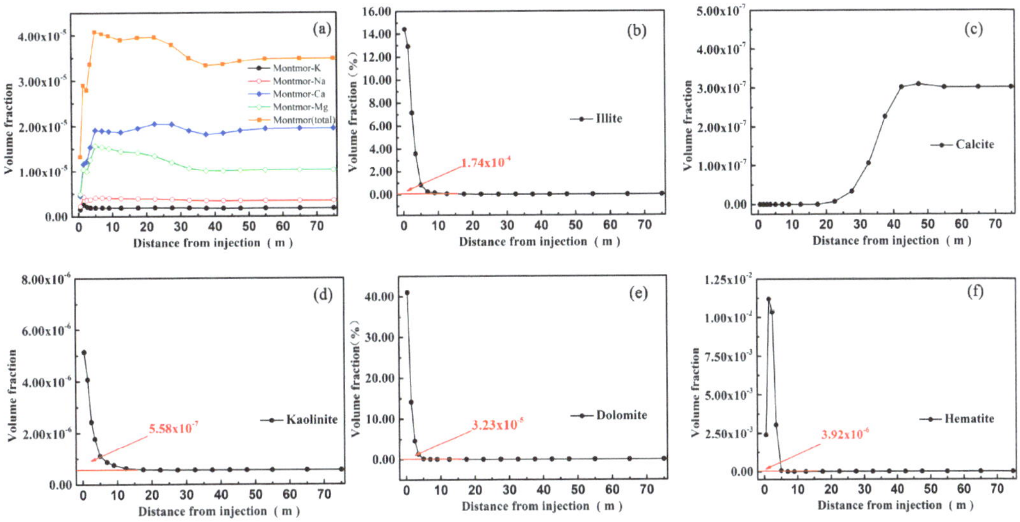

Figure 5. Precipitation of secondary minerals. (**a**)—Montmorillonite; (**b**)—Illite; (**c**)—Calcite; (**d**)—Kaolinite; (**e**)—Dolomite; (**f**)—Hematite.

In sum, significant changes in the porosity and permeability of the near-well reservoir were mainly caused by the dissolution of the initial minerals (K-feldspar, albite, and chlorite) and the secondary precipitation of dolomite and illite. In fact, secondary minerals were generated in the whole reservoir, but they had no significant impact on the reservoir permeability far away from the injection well due to their very low contents. The production of montmorillonite (Equation (13)), calcite (Equation (14)), and kaolinite (Equation (15)) was all below 1×10^{-4}. By contrast, the production of hematite (Equation (16)) was slightly higher, but its highest value near the injection point was only 1.1% of the initial content. Figure 5a shows the total amount of produced montmorillonite and the amounts of the four types of montmorillonite produced. According to this figure, the secondary montmorillonite mainly included calcium montmorillonite and magnesium montmorillonite. The production of montmorillonite and calcite near the injection point was lower than that in the far-well zones. This result is different from the changing trend that the production of other secondary minerals decreased with an increase in the distance from the injection point. This phenomenon occurred because Ca^{2+}, Mg^{2+}, and HCO_3^- were preferentially consumed to form dolomite and illite but the fluid flow and solute transport caused by the injection inhibited the precipitation of montmorillonite and calcite.

$$H^+ + Mg^{2+} + H_2O + \left(K^+,\ Ca^{2+},\ Na^+\right) + SiO_2(aq) + AlO_{2^-} \rightarrow \text{Montmorillonite} \quad (13)$$

$$Ca^{2+} + CO_3{}^{2-} \rightarrow CaCO_3 \quad (14)$$

$$H^+ + H_2O + SiO_2(aq) + AlO_{2^-} \rightarrow \text{Kaolinite} \quad (15)$$

$$Fe^{2+} + H_2O + O_2 \rightarrow Fe_2O_3 + H^+ \quad (16)$$

During the actual operation of the EGS, the process of particle migration (the precipitation of both primary and secondary minerals) caused by fluid injection, as well as the spatial heterogeneity of fracture distribution and scale, may cause suspended particle plugging and distant reservoir. This study simplified the actual conditions without considering this process. However, it is significant to improve the calculation code to include this process in the subsequent work.

5.2. Changes at Injection Point

As concluded from the analysis of the global monitoring results at 10 a of the model, the mineral dissolution and precipitation and the significant changes in the reservoir porosity and permeability mainly occurred near the injection point because of the drastic changes in temperature, flow field, and hydrochemical composition caused by the injection of cold water. Therefore, it is necessary to monitor and analyze the reservoir porosity and permeability at the injection grid, as well as the changes in the mineral dissolution and precipitation with time during the injection. The results are shown in Figure 6.

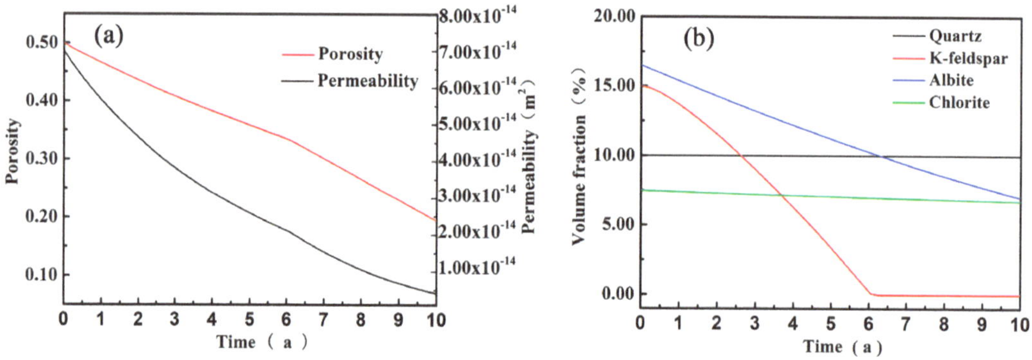

Figure 6. Time-varying simulation results at the injection point. (**a**)—Porosity and permeability; (**b**)—Primary minerals.

As shown in Figure 6, the reservoir porosity and permeability at the injection point decreased continuously as the injection continued, even at 10 a. The injected water yielded the most significant and rapid effect on the K-feldspar dissolution. The effect was intensified as the injection continued, and the K-feldspar completely dissolved by 6 a. The albite and chlorite contents decreased at stable rates, and the chlorite content decreased gently. As shown in Figure 4b, the injected water was more conducive to the dissolution of feldspar minerals, especially K-feldspar.

Figure 7 shows the formation of various secondary minerals at the injection point during the whole operation of the EGS (no calcite was formed for the reasons discussed above, and therefore no calcite is shown in these figures). As shown in Figure 7, the montmorillonite and kaolinite contents only increased continuously in the first 6 months and then were almost stabilized. Their final precipitate quantities were only approximately 1.28×10^{-5} and 5.10×10^{-6}, respectively, which cannot affect the reservoir porosity and permeability. The illite, dolomite, and hematite contents continuously increased during the whole operation of the EGS, and the precipitate quantity of hematite was small and was only 2.40×10^{-3} at 10 a. The generation rate of illite decreased significantly after 6 a. This occurred because the primary K-feldspar was almost all dissolved near the injection point with the continuous progress of fluid injection and solute transport (Figures 4b and 6b), and thus the main K^+ source for the illite formation was lost. According to the analysis in Section 5.1, the illite precipitation was mainly controlled by the dissolution of feldspar minerals, especially K-feldspar. This conclusion is consistent with the analysis results of Xue et al. [37]. Based on the relevant literature, the transformation of K^+ from K-feldspar to illite can be summarized as Equation (17). At 10 a, the illite content at the injection point reached 14.4%.

$$KAlSi_3O_8 + H_2O + Mg^{2+} + H^+ \rightarrow \text{Illite} \tag{17}$$

The dolomite precipitation in the whole injection process (the dolomite content reached 41.0% at 10 a) was the main reason for the significant decrease in the reservoir porosity and permeability at the injection point. The continuous and rapid dolomite formation was attributable to the continuous injection of fluids containing Ca^{2+}, Mg^{2+}, and HCO_3^-. Moreover, the dissolution of the initial mineral chlorite can also provide sufficient Mg^{2+}.

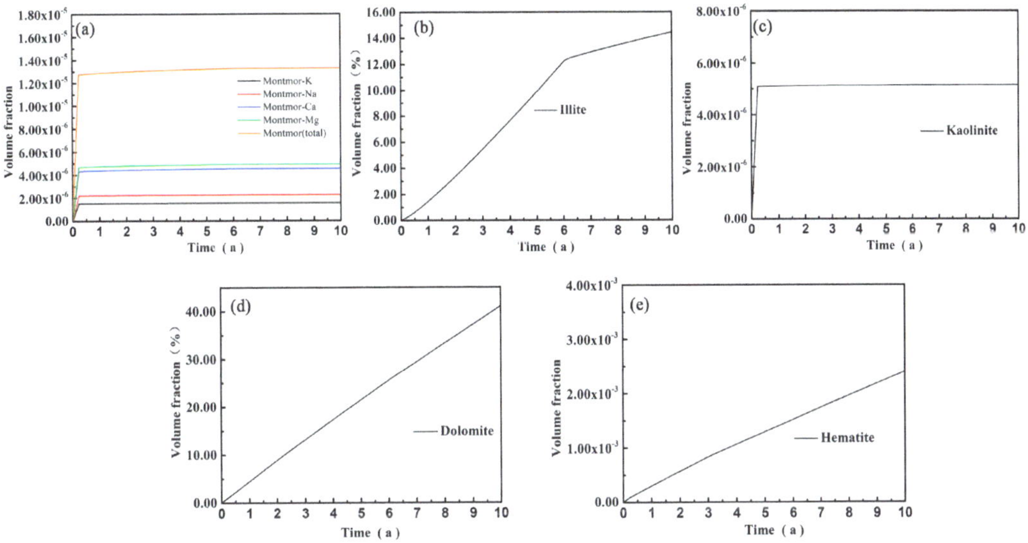

Figure 7. Precipitation of secondary minerals at the injection point. (**a**)—Montmorillonite; (**b**)—Illite; (**c**)—Kaolinite; (**d**)—Dolomite; (**e**)—Hematite.

Overall, the water–rock interactions in the fluid injection process led to the dissolution of primary minerals and the formation of massive secondary minerals, leading to a continuous decrease in the permeability of the near-well reservoir. The intuitive effect of this process on the EGS operation was mainly reflected in the continuous increase in the injection pressure. To further explore the process, the simulation time of the basic scheme was set to 30 a, and the changes in the permeability at the injection grid and fluid pressure over time were monitored. Considering that the plugging process of the reservoir in this study is affected by the selected calculation method for the change in permeability, the calculation method in the original case was changed from the cubic law to the simplified Carman–Kozeny relation (18), and the simulation results (Figure 8) were compared with the original results to further analyze the effect.

$$k = k_i \left(\frac{\varphi}{\varphi i} \right)^3 \frac{(1-\varphi i)^2}{(1-\varphi)^2} \tag{18}$$

As shown in Figure 8, the fluid pressure at the injection grid gradually increased in the long-term injection process with a decrease in the permeability. The fluid pressure increased slowly and stably within the first 10 a and increased sharply after 10 a. At about 12 a, the reservoir permeability at the injection grid was lower than 3.00×10^{-16} m^2 and the fluid pressure rose rapidly in a short time. Therefore, it is considered that the injection grid was plugged at this time. In the actual production process, once the injection pressure exceeds the threshold that the equipment withstands, the injection cannot be sustained. Considering that the operating life of an EGS is generally expected to be 20–30 a, it is necessary to reasonably control the injection flow rate, injection temperature, and hydrochemical composition during the long-term operation and regularly perform chemical treatments to remove the well plugging, aiming to mitigate the damage to the permeability of the near-well reservoir caused by water–rock interactions. Compared with the case using the cubic law, the simulation results obtained using the Carman–Kozeny relation showed a certain gap in the changes in the permeability and fluid pressure at the injection grid. Specifically, the permeability was lower than that of the original plan during the whole injection period. The differences in fluid pressure were not apparent in the early stage (about 7 years) and then gradually increased, with the growth rate increasing sharply. Therefore, in this case, plugging occurred earlier than that in the original scheme (less than 1 a). The possible reason is that the reservoir permeability is more sensitive to porosity changes in the Carman–Kozeny relation, which should be more applicable to porous reservoirs with low permeability. In conclusion, the use of different relations for the change in reservoir permeability has a certain impact on the evolutionary process of plugging but does not change the occurrence and form of plugging.

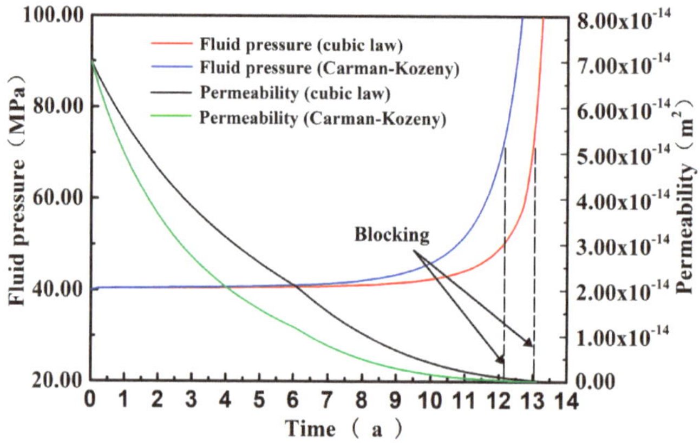

Figure 8. Changes in the permeability and pressure at the injection grid.

5.3. Sensitivity Analysis

5.3.1. Influence of Different Injection Flow Rates

Figure 9 shows the time-varying changes in the porosity of the injection grid, the initial mineral contents, and the contents of main secondary minerals under different injection flow rates and an injection temperature of 70 °C. As the injection flow rate increased, the porosity at the injection point decreased more significantly during the same operation time for two reasons. First, a higher injection flow rate destructed the chemistry equilibrium of the near-well water more significantly, driving the chemical reactions to develop toward the formation of secondary minerals. Second, a higher injection flow rate intensified the dissolution of primary minerals, providing more sufficient material sources for the formation of secondary minerals. The different effects of different injection flow rates on the initial mineral contents were mainly reflected in the K-feldspar dissolution. As shown in Figure 9b, K-feldspar dissolved more rapidly and reached equilibrium in a shorter time under a higher flow rate, leading to a higher formation rate of illite in the early operation stage of the EGS (when K-feldspar was not completely dissolved; Figure 9d). The illite content did not change significantly in the late operation stage of the EGS when the injection flow rate was set to below 5 m^3/h and only increased when the flow rate reached 6 m^3/h. In conclusion, the illite formation was mainly controlled by the K-feldspar dissolution. Both the formation rate and total quantity of dolomite increased significantly with an increase in the injection flow rate, and the formation rates of dolomite under different schemes tended to be stable. These results reflect that the Mg^{2+} content in the system was sufficiently high and the dolomite precipitation was mainly controlled by the injected Ca^{2+} and HCO_3^-.

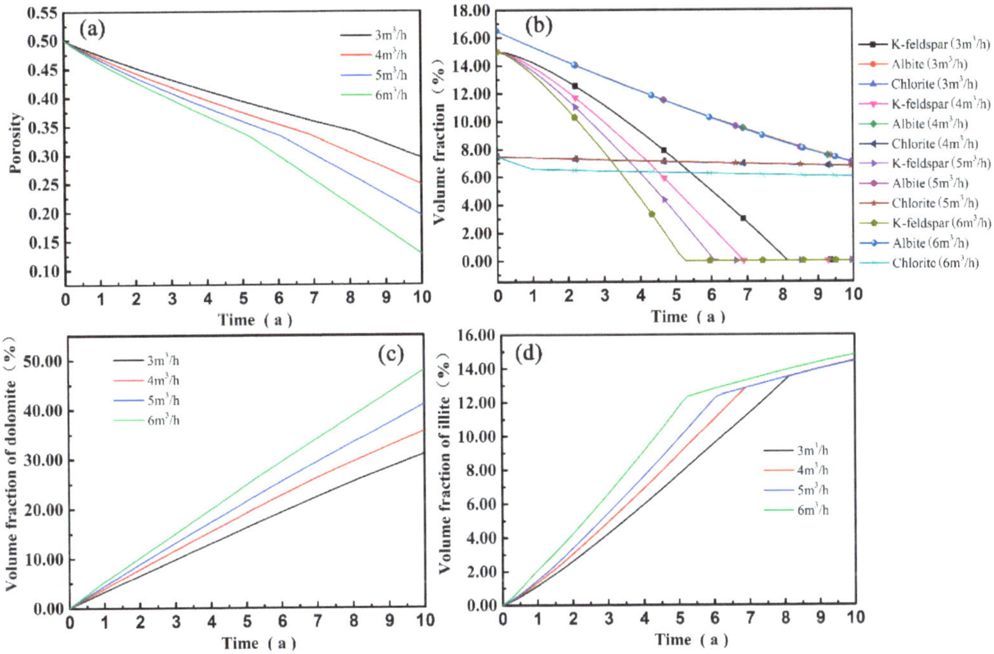

Figure 9. Simulation results at different injection flow rates. (**a**)—Porosity; (**b**)—Permeability; (**c**)—Dolomite; (**d**)—Illite.

5.3.2. Influence of Different Injection Temperatures

Figure 10 shows the time-varying changes in the porosity of the injection grid, the initial mineral contents, and the contents of main secondary minerals under different injection temperatures and an injection flow rate of 5 m^3/h. During the initial injection within the first 3 a, there was no significant difference in the porosity at the injection point under different injection temperatures. The porosity at the injection point during the first 3–5.4 a under an injection temperature of 90 °C was significantly higher than that of the other two schemes. This occurred because the increase in the injection temperature expedited the initial mineral dissolution, effectively counteracting the plugging effect of secondary mineral formation in the near-well reservoir. As shown in Figure 10b, the dissolution rates of K-feldspar and albite significantly increased with an increase in the injection temperature, while the dissolution rate of chlorite only slightly changed when the injection temperature increased from 50 °C to 70 °C and then increased significantly when the injection temperature increased to 90 °C. The differences in porosity between different schemes increased gradually in the late injection stage, and the porosity at the injection point decreased with an increase in the injection temperature. The reason for this trend is that with a decrease in initial mineral contents in the late injection stage, the increase in the reservoir porosity caused by initial mineral dissolution gradually weakened, but secondary minerals were formed continuously. As shown in Figure 10c,d, the formation quantities and rates of dolomite and illite both increased with an increase in the injection temperature, while the formation rate of illite decreased with an increase in the injection temperature after the turning point (note: the turning point appeared increasingly earlier with an increase in the injection temperature, and there was a significant correlation between its appearance time and the K-feldspar dissolution). This occurred because the main K^+ source was lost after the complete dissolution of K-feldspar and more Mg^{2+} was consumed by the dolomite precipitation.

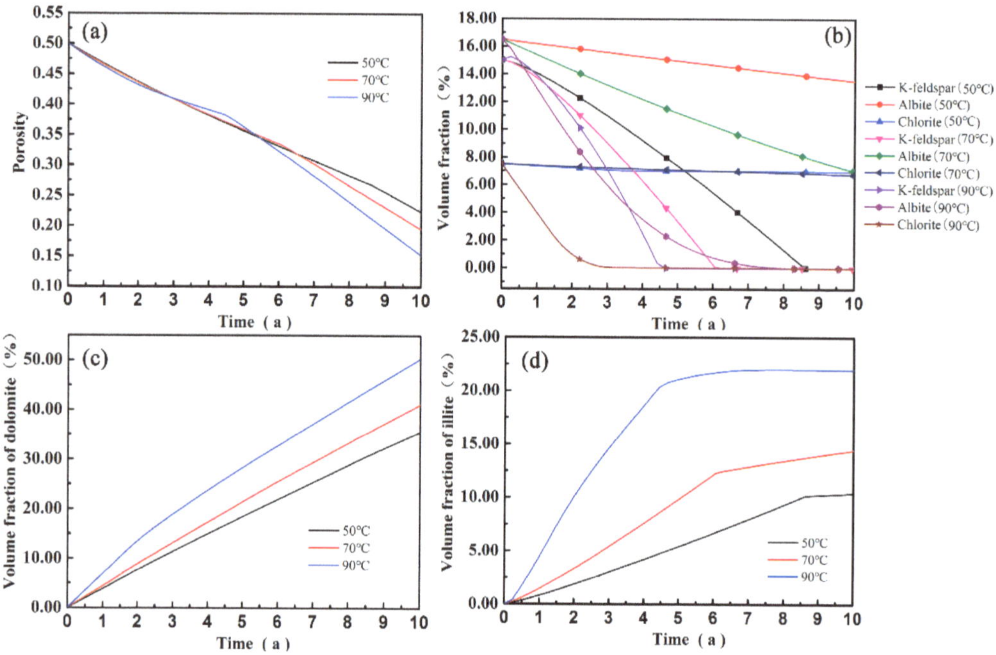

Figure 10. Simulation results at different injection temperatures. (**a**)—Porosity; (**b**)—Permeability; (**c**)—Dolomite; (**d**)—Illite.

5.3.3. Influence of Different Hydrochemical Compositions

Figure 11 shows the simulation results of different chemical composition schemes of injected water under an injection flow rate of 5 m^3/h and an injection temperature of 70 °C. Compared to the basic scheme, the porosity at the injection point in different chemical composition schemes decreased to different degrees. Among them, the porosity in Scheme 8 (Ca^{2+} was the leading cation) decreased slightly, while the porosity in Scheme 7 (K^+ was the leading cation) decreased to below 0.05 at 7a. In this case, the fluid pressure at the injection grid was too high to ensure continuous and stable injection. In Scheme 9 (Mg^{2+} was the leading cation), a substantial decrease in the porosity appeared at 4a. Therefore, it can be preliminarily determined that Mg^{2+} had the most significant influence on the porosity and permeability of the near-well reservoir in the study area.

Changes in the chemical composition of injected water have different degrees of effects on the dissolution of initial minerals and the generation of secondary minerals. The dissolution of initial minerals in Scheme 7 was most significantly different from that in the basic scheme. The high increase in the K^+ concentration in this scheme inhibited the dissolution of primary K-feldspar and led to the formation of massive secondary K-feldspar (Figure 11b). This process and the sharp decrease in the Na^+ concentration greatly promoted the albite and chlorite dissolution (Figure 11c,d), thus providing more sufficient Mg^{2+} for the dolomite formation (Figure 11e). In addition, since no K-feldspar dissolved in Scheme 7 and the K^+ in the injected fluid was mainly consumed by the formation of secondary K-feldspar, the illite formation was inhibited and the precipitate quantity of illite was only 3.96×10^{-4} at 7a. The influence of the increase in the Mg^{2+} concentration on the mineral contents in Scheme 9 was mainly reflected in the formation of secondary minerals. As shown in Figure 11e,f, the generation rates of dolomite and illite significantly increased compared with other schemes. In Scheme 9, as illite continued to be formed and the chlorite dissolution gradually reached equilibrium in the late injection stage, the

K-feldspar dissolution intensified, releasing a large amount of K^+ and Mg^{2+} and further increasing the formation rate of the two main secondary minerals, i.e., illite and chlorite.

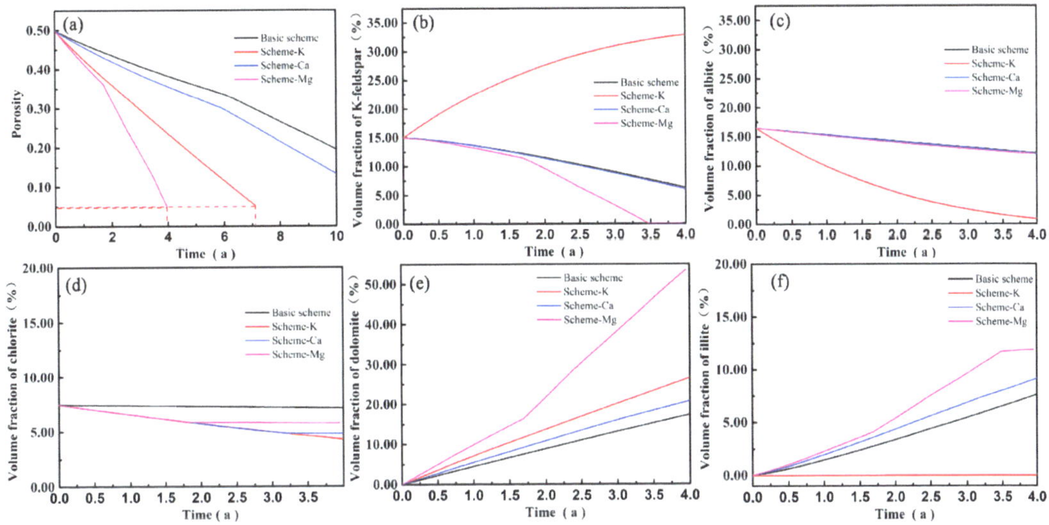

Figure 11. Simulation results under different hydrochemical compositions. (**a**)—Porosity; (**b**)—K-Feldspar; (**c**)—Albite; (**d**)—Chlorite; (**e**)—Dolomite; (**f**)—Illite.

In sum, Mg^{2+} and K^+ have the most significant effects on mineral dissolution and precipitation and the permeability of the near-well reservoir among the chemical components in injected water. Therefore, it is necessary to control the Mg^{2+} and K^+ contents by adding metal ion chelating agents and corrosion inhibitors to the injected water in the long-term operation of geothermal wells.

6. Conclusions

Based on the investigation of Well M-1 in the EGS in the Matouying uplift of Hebei Province, China, this study established a THC reactive solute transport model using the TOUGHREACT program. Using this model, this study explored the mineral dissolution and precipitation laws of the near-well reservoir and their influencing mechanisms on reservoir porosity and permeability in the long-term injection process of this well and analyzed the effects of different injection conditions (injection flow rate, injection temperature, and hydrochemical composition) on the injection process. The conclusions are as follows.

(1) A significant decrease in the porosity and permeability of the EGS reservoir in the Matouying uplift occurred 0–7.5 m away from Well M-1 in the long-term injection process. At 10 a, the porosity at the injection point decreased by 60.0% and the permeability decreased by 94.2%. This occurred due to the dissolution of primary K-feldspar, albite, and chlorite and the formation of secondary minerals dominated by dolomite and illite.

(2) The decreasing rates of the permeability and the porosity of the near-well reservoir in the EGS of the Matouying uplift increased with an increase in the injection flow rate. The reason is that a higher injection flow rate expedited the dissolution of primary K-feldspar, thus providing more sufficient ions for the formation of secondary minerals and promoting the formation of dolomite and illite.

(3) In this simulation, the change in the injection temperature had significant effects on the dissolution of primary minerals and the precipitation of main secondary minerals. Moreover, it had a slight influence on the variation trends of the porosity and permeability of the near-well reservoir in the early injection stage. However, with a decrease in the primary mineral content and the continuous formation of secondary minerals, the decreasing

rates of the porosity and permeability of the near-well reservoir significantly increased with an increase in the injection temperature.

(4) The increase in the K^+ and Mg^{2+} contents in the main chemical components of the injected water expedited the precipitation of dolomite and illite and further greatly exacerbated the plugging of the near-well reservoir in the EGS of the Matouying uplift. Therefore, to improve the operating life of the EGS, it is absolutely necessary to control the content of metal ions, especially K^+ and Mg^{2+}, in the injected water during the long-term injection.

Author Contributions: Conceptual model and numerical simulation, B.F. and Z.C.; visualization and writing, X.L.; sample collection and data acquisition, S.S. and X.Q.; funding acquisition and resource administration, S.L. All authors have read and agreed to the published version of the manuscript.

Funding: This work was supported by the National Key R&D Program of China (No. 2018YFB1501802), the S&T Program of Hebei Provincial (No. 20374201D), the Chinese Academy of Geological Sciences Fund (No. SK202104), the Jilin Provincial Department of Science and Technology (No. 20200403147SF), the Academician Workstation and Academician Cooperation Key Unit Construction of the Second Geological Team of Hebei Coal Geology Bureau (No. 205A4101H), and the research and application demonstration of single-well heat recovery technology and ground comprehensive heat utilization technology for the thermal energy of dry hot rocks (No. 19274102D).

Institutional Review Board Statement: Not applicable.

Informed Consent Statement: Not applicable.

Data Availability Statement: The original contributions presented in the study are included in the article, and further inquiries can be directed to the corresponding author.

Acknowledgments: The authors would like to extend their sincere gratitude to the relevant institutions for financial and technical support.

Conflicts of Interest: The authors declare no conflict of interest.

References

1. Soltani, M.; Kashkooli, F.M.; Dehghani-Sanij, A.R.; Kazemi, A.R.; Bordbar, N.; Farshchi, M.J.; Elmi, M.; Gharali, K.; Dusseault, M.B. A comprehensive study of geothermal heating and cooling systems. *Sustain. Cities Soc.* **2019**, *44*, 793–818. [CrossRef]
2. Hu, X.; Lv, J.; Li, S.; Du, G.; Wang, Z.; Li, H.; Zhu, H. Joint interpretation technology of favorable HDR geothermal resource exploration in Northern Songliao Basin. *Unconv. Resour.* **2022**, *2*, 133–138. [CrossRef]
3. Rathnaweera, T.D.; Wu, W.; Ji, Y.; Gamage, R.P. Understanding injection-induced seismicity in enhanced geothermal systems: The coupled thermo-hydro-mechanical-chemical process to anthropogenic earthquake prediction. *Earth-Sci. Rev.* **2020**, *205*, 103182. [CrossRef]
4. Wang, D.; Dong, Y.; Li, Y.; Wang, Y.; Li, Y.; Liu, H.; Zhang, W.; Sun, D.; Yu, B. Numerical simulation of heat recovery potential of hot dry rock under alternate temperature loading. *Unconv. Resour.* **2022**, *2*, 170–182. [CrossRef]
5. Kharseh, M.; Al-Khawaja, M. Optimizing Geothermal Energy Resource. *Mech. Eng. Autom.* **2015**, *5*, 667–675.
6. Barcelona, H.; Senger, M.; Yagupsky, D. Resource assessment of the Copahue geothermal field. *Geothermics* **2021**, *90*, 101987. [CrossRef]
7. El Bouazouli, A.; Baidder, L.; Diagana, O.; Rhouzlane, S.; Bichri, A. Prospecting of potential geothermal resources from the east to the Moroccan Sahara. *Mater. Today Proc.* **2020**, *27*, 3224–3233. [CrossRef]
8. Olasolo, P.; Juárez, M.C.; Morales, M.P.; D'Amico, S.; Liarte, I.A. Enhanced geothermal systems (EGS): A review. *Renew. Sustain. Energy Rev.* **2016**, *56*, 133–144. [CrossRef]
9. Alegria, P.; Catalan, L.; Araiz, M.; Rodriguez, A.; Astrain, D. Experimental development of a novel thermoelectric generator without moving parts to harness shallow hot dry rock fields. *Appl. Therm. Eng.* **2022**, *200*, 117619. [CrossRef]
10. Lv, Y.; Yuan, C.; Zhu, X.; Gan, Q.; Li, H. THMD analysis of fluid injection-induced fault reactivation and slip in EGS. *Geothermics* **2022**, *99*, 102303. [CrossRef]
11. Asai, P.; Podgorney, R.; McLennan, J.; Deo, M.; Moore, J. Analytical model for fluid flow distribution in an Enhanced Geothermal Systems (EGS). *Renew. Energy* **2022**, *193*, 821–831. [CrossRef]
12. Borgia, A.; Pruess, K.; Kneafsey, T.J.; Oldenburg, C.M.; Pan, L. Simulation of CO_2-EGS in a Fractured Reservoir with Salt Precipitation. *Energy Procedia* **2013**, *37*, 6617–6624. [CrossRef]
13. Driba, D. 1D Thermal-Hydraulic-Chemical (THC) Reactive transport modeling for deep geothermal systems: A case study of Groß Schönebeck reservoir, Germany. In *Agu Fall Meeting*; American Geophysical Union: San Francisco, CA, USA, 2014.
14. Ren, G.; Jiang, J.; Younis, R.M. A Model for Coupled Geomechanics and Multiphase Flow in Fractured Porous Media Using Embedded Meshes. *Adv. Water Resour.* **2018**, *122*, 113–130. [CrossRef]

15. Salimzadeh, S.; Paluszny, A.; Nick, H.M.; Zimmerman, R.W. A Three-Dimensional Coupled Thermo-Hydro-Mechanical Model for Deformable Fractured Geothermal Systems. *Geothermics* **2017**, *71*, 212–224. [CrossRef]
16. Pei, Y.; Sepehrnoori, K. Efficient Modeling of Depletion Induced Fracture Deformation in Unconventional Reservoirs. In Proceedings of the SPE Annual Technical Conference and Exhibition, Dubai, United Arab Emirates, 21–23 September 2021. [CrossRef]
17. Li, T.; Han, D.; Yang, F.; Li, J.; Wang, D.; Yu, B.; Wei, J. Modeling Study of the Thermal-Hydraulic-Mechanical Coupling Process for EGS Based on the Framework of EDFM and XFEM. *Geothermics* **2022**, *89*, 101953. [CrossRef]
18. Yuan, Y.; Xu, T.; Moore, J.; Lei, H.; Feng, B. Coupled Thermo–Hydro–Mechanical Modeling of Hydro-Shearing Stimulation in an Enhanced Geothermal System in the Raft River Geothermal Field, USA. *Rock Mech. Rock Eng.* **2020**, *53*, 5371–5388. [CrossRef]
19. Regenspurg, S.; Feldbusch, E.; Byrne, J.; Deon, F.; Driba, D.L.; Henninges, J.; Kappler, A.; Naumann, R.; Reinsch, T.; Schubert, C. Mineral precipitation during production of geothermal fluid from a Permian Rotliegend reservoir. *Geothermics* **2015**, *54*, 122–135. [CrossRef]
20. Ma, Z.; Xu, Y.; Zhai, M.; Wu, M. Clogging mechanism in the process of reinjection of used geothermal water: A simulation research on Xianyang No.2 reinjection well in a super-deep and porous geothermal reservoir. *J. Groundw. Sci. Eng.* **2017**, *5*, 311–325.
21. Yanaze, T.; Yoo, S.; Marumo, K.; Ueda, A. Prediction of permeability reduction due to silica scale deposition with a geochemical clogging model at Sumikawa Geothermal Power Plant. *Geothermics* **2019**, *79*, 114–128. [CrossRef]
22. Zhao, Y.; Feng, B.; Zhang, G.; Shangguan, S.; Qi, X.; Li, X.; Qiao, Y.; Xu, J. Study of the interaction between the granitic hot-dry rock (HDR) and different injection waters. *Acta Geol. Sin.* **2020**, *94*, 2115–2123.
23. Ke, Z.; Feng, B.; Liu, Y.; Cui, Z.; Liu, X. Dissolution and sedimentation patterns of typical minerals in artificial reservoirs under different environment. *Unconv. Resour.* **2022**, *2*, 60–71. [CrossRef]
24. Feng, C.; Gao, G.; Zhang, S.; Sun, D.; Zhu, S.; Tan, C.; Ma, X. Fault slip potential induced by fluid injection in the Matouying enhanced geothermal system (EGS) field, Tangshan seismic region, North China. *Nat. Hazards Earth Syst. Sci.* **2022**, *22*, 2257–2287. [CrossRef]
25. Qi, X.; Shangguan, S.; Zhang, G.; Pan, M.; Su, Y.; Tian, L.; Li, X.; Qiao, Y.; Zhang, J. Site selection and developmental prospect of a hot dry rock resource project in the Matouying Uplift, Hebei Province. *Earth Sci. Front.* **2020**, *27*, 94–102.
26. Shangguan, S. Occurrence conditions of hot-dry-rock geothermal resources and development prospects in Matouying area. *China Energy Environ. Prot.* **2017**, *39*, 155–165.
27. Zhang, Y.; Gao, L.; Liu, X.; Zhao, Y.; Wang, D.; Zhang, J. Drilling technology of the M-1 well in hot dry rock of Matouying, Tangshan. *Geol. Explor.* **2022**, *58*, 176–186.
28. Xu, T.F.; Spycher, N.; Sonnenthal, E.; Zhang, G.X.; Zheng, L.E.; Pruess, K. TOUGHREACT Version 2.0: A simulator for subsurface reactive transport under non-isothermal multiphase flow conditions. *Comput. Geosci.* **2011**, *37*, 763–774. [CrossRef]
29. Xu, T.; Sonnenthal, E.; Spycher, N.; Pruess, K. *TOUGHREACT User's Guide: A Simulation Program for Non-Isothermal Multiphase Reactive Geochemical Transport in Variably Saturated Geologic Media*; V1.2.1; Lawrence Berkeley National Lab. (LBNL): Berkeley, CA, USA, 2008.
30. Xu, T.; Hu, Z.; Feng, B.; Feng, G.; Li, F.; Jiang, Z. Numerical evaluation of building heating potential from a co-axial closed-loop geothermal system using wellbore–reservoir coupling numerical model. *Energy Explor. Exploit.* **2020**, *38*, 733–754. [CrossRef]
31. Qi, X.; Zhang, G.; Shangguan, S.; Su, Y.; Tian, L.; Li, X.; Qiao, Y.; Liu, X. A brief analysis of hot and dry rock geothermal resource hosting and distribution in Hebei Province. *Coal Geol. China* **2018**, *30*, 47–73.
32. Cui, Z.; Shangguan, S.; Gherardi, F.; Qi, X.; Xu, J.; He, S.; Feng, B. Experimental Study on the Effect and Mechanism of Chemical Stimulation on Deep High-Temperature Granite. *Front. Earth Sci.* **2022**, *10*, 893969. [CrossRef]
33. Shangguan, S.; Sun, D.; Zhang, G.; Yang, Y.; Qi, X.; Chen, D.; Qiao, Y.; Li, A.; Chen, Q. In-situ stress measurement and fault stability analysis within a depth of 3–4 km in the Tangshan area. *Acta Geol. Sin.* **2021**, *95*, 3915–3925.
34. Xu, J.; Feng, B.; Cui, Z.; Liu, X.; Ke, Z.; Feng, G. Comparative Study of Acid and Alkaline Stimulants with Granite in an Enhanced Geothermal System. *Acta Geol. Sin.* **2021**, *95*, 1926–1939. [CrossRef]
35. Na, J.; Xu, T.F.; Jiang, Z.J.; Bao, X.H.; Wu, Y.D.; Feng, B. A study on the interaction of mud acid with rock for chemical stimulation in an enhanced geothermal system. *Environ. Earth Sci.* **2016**, *75*, 1025. [CrossRef]
36. Yang, Z.; Wang, F.; Yang, B.; Tian, H.; Xu, T. Numerical simulation of the influence of Chlorite on the content of mineral trapping of CO_2 in the sandstone. *Bull. Mineral. Petrol. Geochem.* **2014**, *33*, 201–207.
37. Xue, W.; Tan, X.; Li, Z.; Qiu, Y.; Yong, Y.; Jiang, W. Dissolution mechanism of feldspars in the clastic rocks and its contribution to diagenesis. *Complex Hydrocarb. Reserv.* **2015**, *1*, 1–6, 61.

energies

MDPI

Article

A Hybrid Oil Production Prediction Model Based on Artificial Intelligence Technology

Xiangming Kong *, Yuetian Liu *, Liang Xue, Guanlin Li and Dongdong Zhu

State Key Laboratory of Petroleum Resources and Prospecting, China University of Petroleum (Beijing), Beijing 102249, China

* Correspondence: kongxm@student.cup.edu.cn (X.K.); lyt51x@163.com (Y.L.)

Abstract: Oil production prediction plays a significant role in designing programs for hydrocarbon reservoir development, adjusting production operations and making decisions. The prediction accuracy of oil production based on single methods is limited since more and more unconventional reservoirs are being exploited. Artificial intelligence technology and data decomposition are widely implemented in multi-step forecasting strategies. In this study, a hybrid prediction model was proposed based on two-stage decomposition, sample entropy reconstruction and long short-term memory neural network (LSTM) forecasts. The original oil production data were decomposed into several intrinsic mode functions (IMFs) by complete ensemble empirical mode decomposition with adaptive noise (CEEMDAN); then these IMFs with different sample entropy (SE) values were reconstructed based on subsequence reconstruction rules that determine the appropriate reconstruction numbers and modes. Following that, the highest-frequency reconstructed IMF was preferred to be decomposed again by variational mode decomposition (VMD), and subsequences of the secondary decomposition and the remaining reconstructed IMFs were fed into the corresponding LSTM predictors based on a hybrid architecture for forecasting. Finally, the prediction values of each subseries were integrated to achieve the result. The proposed model makes predictions for the well production rate of the JinLong volcanic reservoir, and comparative experiments show that it has higher forecasting accuracy than other methods, making it recognized as a potential approach for evaluating reservoirs and guiding oilfield management.

Keywords: two-stage decomposition; sample entropy; hybrid model; time series forecasting; oil production forecast

1. Introduction

Well production is one of the most important indicators of oilfield development and management. Acquainting well production performance in advance can help engineers adapt development countermeasures and optimize development effects timely. Decline curve analysis has been widely utilized and achieves a good performance in conventional reservoirs [1]. The Arps model, however, may not be suitable due to the intricacy of flow dynamics in unconventional reservoirs. Under certain assumptions, the formation parameters are simplified, and the analytical or semi-analytical model is proposed and solved [2,3], which can simplify the complex formation seepage issue, but also limit the model's application. Numerical reservoir simulation techniques make production forecasts based on history matching by building a geological model of the actual reservoir [4]; however, establishing a model that is virtually identical to the actual reservoir requires reservoir engineers to have considerable experience. The complex geological characteristics of unconventional reservoirs, on the other hand, exacerbate the non-linear variance of oil production over time, making production prediction extremely challenging.

Various artificial intelligence algorithms have been implemented in the field of petroleum engineering with the growth of machine learning theory, paving a new route for the investigation of the production prediction issue [5,6]. Wang constructed a deep neural network

(DNN) model to forecast cumulative oil production of Bakken shale reservoirs [7]. On the basis of a long short-term memory (LSTM) structure, Huang conducted the development prediction task in a water-flooding reservoir [8]. Sagheer and Kotb established a deep long-short term memory (DLSTM) framework to enhance the oil production forecast performance and employed the genetic algorithm to optimize the hyperparameters [9]. Cheng used the long short-term memory (LSTM) network and gated recurrent unit (GRU) method to predict the oil production of actual oilfields in China and India [10]; the results indicate that LSTM and GRU have respective advantages under different circumstances.

Whereas single models are not sufficiently applicable for complicated issues, hybrid structures have become a research trend in the time series forecasting area [11,12], including well performance forecasting [13]. Fan developed a hybrid model that incorporated the autoregressive integrated moving average (ARIMA) with the long short-term memory (LSTM) network to predict the production of three actual wells under the influence of manual operations [14]. Li [15] used the PSO algorithm to optimize the proposed CNN-LSTM production forecast model, which has higher prediction accuracy than a single model. To enhance prediction validity, a current trend in time series forecasting is to incorporate the artificial intelligence algorithm with decomposition pre-processing strategies [16–18]. Liu proposed a hybrid model that combines ensemble empirical mode decomposition (EEMD) with an LSTM network [19], with the appropriate intrinsic mode functions (IMFs) of EEMD chosen using dynamic time warping (DTW). The method achieved a higher accuracy than other models in two reservoirs. Wang constructed a hybrid method with variational mode decomposition (VMD) and gated recurrent unit (GRU) [20], which was implemented in the Tahe oilfield and demonstrated an outstanding performance.

Oil production data of unconventional reservoirs are complicated and nonstationary; even a decomposition-forecasting method could not obtain excellent accuracy, so further data processing [21] is necessary. By considering both improving the prediction performance and reducing the accumulation errors, the secondary decomposition method and subsequence reconstruction approach are implemented comprehensively in this study. By analyzing the prediction performance of various reconstruction numbers and modes, guidelines for component reconstruction are generated, and the threshold is set for sample entropy values of first-stage decomposition IMFs. After evaluating the forecast efficiency, the most complex subsequence is further decomposed by VMD. As the core part of the structure, an optimum predictor is selected among four artificial intelligence algorithms to utilize their excellent data-learning skills. Then, the multi-type subseries data is inputted to the corresponding LSTM predictor based on a hybrid architecture. The multi-stage prediction structure is proposed and applied to forecast the well production rate of the Jinlong volcanic reservoir. The contributions of the proposed method are as follows:

(1) A novel multi-step decomposition-integration framework is established for oil production forecasting;
(2) Intrinsic mode functions (IMFs) are reconstructed to re-IMFs according to the rules for subsequence reconstruction numbers and modes, which reduces accumulation errors and calculation complexity;
(3) The highest-frequency reconstructed IMF is preferred to be further decomposed to enhance prediction accuracy;
(4) The hybrid model combines the advantages of both integral and corresponding architectures, maintaining both prediction accuracy and computing efficiency.

2. Methods

2.1. Complete Ensemble Empirical Mode Decomposition with Adaptive Noise

Empirical Mode Decomposition (EMD) has been popularly utilized for decomposing sequential data in several time series forecasting fields. It also has disadvantages, such as weak stability. Complete Ensemble Empirical Mode Decomposition with Adaptive Noise (CEEMDAN) is presented as an improved practice of EMD [22]. The raw time series data can also be decomposed into several intrinsic mode functions (IMFs) and a residue with different frequencies by the CEEMDAN, which incorporates adaptive noise into the EMD process, and the decomposition process is complete so the least reconstruction error is obtained, helping resolve the issue of modal aliasing and residual noise in the sequence.

2.2. *Variational Mode Decomposition*

Variational Mode Decomposition (VMD) is a novel non-recursive data decomposition algorithm defined by Dragomiretskiy and Zosso [23] to solve the limitations of sensitivity to noise and sampling, which can decompose nonlinear and nonstationary original data into specific amounts of intrinsic mode functions (IMFs). The VMD method searches for the optimal solution to a variational problem to accomplish adaptive decomposition. The promotion points of VMD include minimizing the sum of evaluated bandwidth and inhibiting noise. It has been utilized to further decompose high-frequency subseries data from previous data processing, which can effectively decrease its complexity.

2.3. *Long Short-Term Memory Network*

Recurrent neural network (RNN) is widely utilized in Natural Language Processing (NLP) and Time Series Forecasting (TSF) areas; because of the gradient disappearance and explosion problem, improved methods have been proposed, especially the long short-term memory neural network (LSTM) which demonstrates excellent performance in dealing with many issues [24]. Figure 1 depicts the cell structure of the LSTM, which consists of the cell state, forget gate, input gate and output gate. As the core section of LSTM structure, a cell state contains information about all previous states, and at each new time step, operations are carried out to identify which old information to discard and which new information to add.

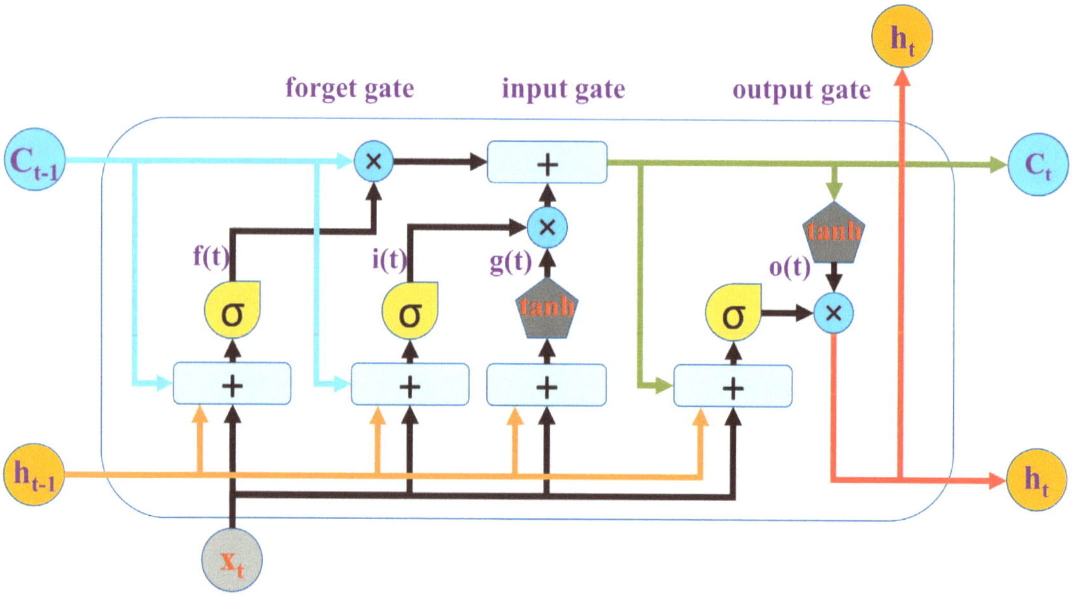

Figure 1. Structure of the LSTM.

2.4. Sample Entropy

Sample entropy (SE), proposed by Richman and Moorman [25], can be employed to evaluate the complexity of time series; the higher the sample entropy value, the more complicated the sequence is. Although the series data is decomposed, IMFs still have several high-frequency subsequences. Based on the rules of reconstruction numbers and modes we defined in Section 2.5, IMFs decomposed from different source data could be integrated in a general flow, which can properly decrease computing workload and promote model efficiency.

2.5. Rules of Subsequence Reconstruction and Secondary Decomposition

To reduce the accumulation errors, simplify the complexity of computation, and further process high-complexity components, this research implements a procedure after initial decomposition with the production data of actual oil wells: reconstruction and secondary decomposition. We determine the optimum number of reconstruction subsequences and prefer the most appropriate mode for the method's wide adaptability and migration. Details of the rules' definition process for this flow are depicted by comparable experiments in Section 3.2. The conclusions of the experiments are summarized as follows:

(1) The proper number of reconstructed IMFs is set to three according to the prediction performance comparison of multiple hybrid models with various reconstructed IMF counts;

(2) Based on the first decomposition process results of oil well production data in the JinLong volcanic reservoir, the corresponding most appropriate reconstruction modes are preferred. The optimum reconstruction modes show the threshold of the sample entropy value of IMFs to conduct the integration. The high-frequency subseries whose sample entropy values are over 1.0 and the low-complexity IMFs with values under 0.2 should be reconstructed, while the rest of the sequences comprise a re-IMF, regardless of whether the initial decomposition's component number is 8 or 9.

(3) To consider both improving the prediction accuracy and simplifying the complexity of calculation, the secondary decomposition is applied to process only the highest frequency subsequence among the three reconstructed IMFs.

2.6. Architecture of the Proposed Hybrid Model

The architecture of the hybrid model proposed in this paper is depicted in Figure 2. The main progress can be described as follows:

Step 1 Collect the actual oil production data.
Step 2 Decompose time series data into several IMFs by CEEMDAN.
Step 3 Calculate the sample entropy values of all IMFs and reconstruct them into fewer re-IMFs based on the rules for component reconstruction numbers and modes.
Step 4 Decompose the highest-frequency re-IMF0 from Step 3 by VMD to obtain new subsequences, and feed them into an integral LSTM architecture in the form of a matrix for prediction.
Step 5 Build the same number of LSTM models as reconstructed IMFs without secondary decomposition, input each IMF vectors and forecast them correspondingly.
Step 6 Integrate all the forecasting values of each re-IMF from Step 4 and Step 5 to obtain the final prediction result and evaluate it.

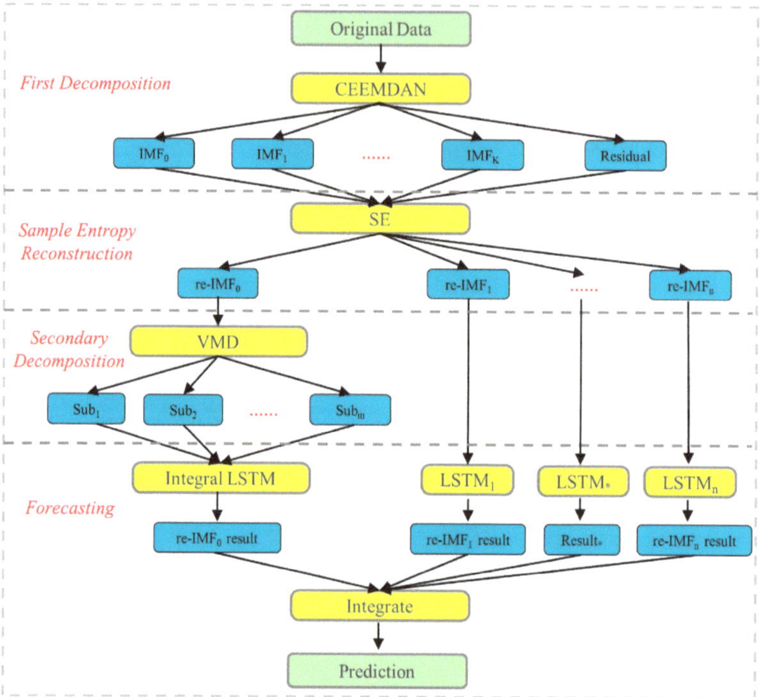

Figure 2. Framework of the proposed hybrid model.

2.7. Model Evaluation Index

For evaluating the forecasting performance of every model, this study selects the following four common performance measurement indices: root mean square error (RMSE), mean absolute error (MAE), mean absolute percentage error (MAPE) and determination coefficient (R^2). The calculation formulas for the indices are shown as follows:

$$\text{RMSE} = \sqrt{\frac{1}{n}\sum_{i=1}^{n}(y_i - \hat{y}_i)^2} \quad (1)$$

$$\text{MAE} = \frac{1}{n}\sum_{i=1}^{n}|y_i - \hat{y}_i| \quad (2)$$

$$\text{MAPE} = \frac{1}{n}\sum_{i=1}^{n}\frac{|y_i - \hat{y}_i|}{y_i} \times 100 \quad (3)$$

$$R^2 = \frac{\sum_{i=1}^{n}(\hat{y}_i - \overline{y}_i)^2}{\sum_{i=1}^{n}(y_i - \overline{y}_i)^2} \quad (4)$$

where y_i, $\hat{y}_i$, $\overline{y}_i$ are the actual, forecast and mean values of the time series sample data, respectively, n is the sample size. The values of the four indexes represent the prediction accuracy of the proposed hybrid model; if the values of RMSE, MAE, MAPE are closer to 0, or R^2 is closer to 1, the model is more excellent.

3. Experiments

Aiming to validate the established hybrid forecasting structure, oil production data from the JinLong volcanic reservoir are processed, and seven comparative experiments are conducted.

3.1. Data Preparation

The hybrid model proposed above is implemented to forecast actual oil production in the JinLong(JL) volcanic reservoir, which is located on the east slope of the Zhongguai Uplift in the southwest margin of the Junggar Basin, as shown in Figure 3. The Jiamahe Formation of the Permian is the primary oil-bearing layer to develop, which has an average depth of 4000 m. The thickness of the volcanic rock of the Jiamuhe Formation ranges from 22 m to 286 m, averaging 145.3 m. As a naturally fractured reservoir, oblique fractures, straight split fractures, reticular fractures, microfractures and partially filled fractures develop in the JL volcanic reservoir. The permeability of the formation varies from 0.01 to 68 mD and the average value is 0.56 mD; the porosity ranges from 8% to 22.3% with an average of 12.35%. Oil production data of the production wells in the JL volcanic reservoir are applied in multiple comparative experiments from Sections 3.3–3.8. After processing zero values, missing and abnormal values, the volume of oil production data for each single well varies from 916 to 1309. For the purpose of a high-quality modeling effect, the last 100 samples of each series of data are chosen as test sets; the rest is for training.

Figure 3. Location of the research area.

3.2. Sample Entropy Reconstruction and Secondary Decomposition

After the CEEMDAN process, oil production data could be decomposed into several IMFs, including a residual series. There are two kinds of first-decomposition results determined by the source data in this study: 8 and 9. The sample entropy values of each subsequence are calculated and listed in Table 1. In order to decrease the computational workload on the premise of ensuring prediction accuracy, we integrate the first-stage IMFs into fewer ones and implement secondary decomposition. The principles of reconstruction and secondary decomposition are defined by the following trials:

Table 1. First-stage decomposition results and sample entropy values of each subsequence.

Subsequence	Sample Entropy Value			
	Well-1	Well-2	Well-3	Well-4
IMF0	1.6896	1.6283	1.1927	1.3437
IMF1	1.3168	1.8620	1.4169	1.3095
IMF2	0.7788	1.2472	1.0695	0.6152
IMF3	0.5569	0.6540	0.6434	0.5737
IMF4	0.3745	0.2106	0.2698	0.4166
IMF5	0.1186	0.1255	0.1168	0.1666
IMF6	0.0442	0.0334	0.0572	0.0835
IMF7	0.0040	0.0277	0.0260	0.0084
IMF8	\	0.0022	0.0001	\

(1) Determine the proper number of reconstruction IMFs.

To optimize the best reconstruction number, comparison experiments with different re-IMF quantities are carried out. All the experiments are based on the decomposition-reconstruction-integral LSTM structure. Table 2 shows the specific programs and evaluation index of these models' forecasting performance.

Table 2. Comparison of prediction models with different reconstructed IMF numbers.

Number of Re-IMFs	Well-1		Well-2		Well-3		Well-4	
	R^2	RMSE	R^2	RMSE	R^2	RMSE	R^2	RMSE
1 (no decomposition)	0.8030	3.4270	0.8044	3.1330	0.7658	0.8804	0.7955	4.3385
2	0.8035	3.1414	0.8064	2.5684	0.7801	0.8532	0.8119	1.9584
3	0.8529	2.2888	0.8656	2.1305	0.8549	0.6930	0.8603	1.3145
4	0.8135	2.7132	0.8064	2.4053	0.8134	0.7404	0.7853	2.4741
8 or 9 (no reconstruction)	0.8229	2.5888	0.8346	2.7063	0.7697	0.8731	0.8146	2.3529

When the number of re-IMFs is three, the prediction errors are the lowest and the R^2 scores are the highest. Therefore, in this research, the number of reconstruction IMFs is defined as three to obtain the best prediction accuracy.

(2) Identify the optimum reconstruction modes.

To obtain three reconstruction subsequences, several optional integration modes of IMFs are listed in Tables 3 and 4. The forecasting results of these modes are compared with the same predictor, and the evaluation results are also shown in Tables 3 and 4.

Table 3. Reconstruction modes (first-stage decomposition IMF's number: 8).

Reconstruction Mode	Component of Re-IMF0	Component of Re-IMF1	Component of Re-IMF2	Well-1		Well-4	
				R^2	RMSE	R^2	RMSE
A	IMF0, IMF1	IMF2, IMF3, IMF4	IMF5, IMF6, IMF7	0.8529	2.2888	0.8603	1.3145
B	IMF0, IMF1, IMF2	IMF3, IMF4	IMF5, IMF6, IMF7	0.8056	3.1018	0.8167	1.8762
C	IMF0, IMF1	IMF2, IMF3	IMF4, IMF5, IMF6, IMF7	0.8115	2.6132	0.8186	1.6709

Table 4. Reconstruction modes (first-stage decomposition IMF's number: 9).

Reconstruction Mode	Component of Re-IMF0	Component of Re-IMF1	Component of Re-IMF2	Well-2		Well-3	
				R^2	RMSE	R^2	RMSE
A	IMF0, IMF1, IMF2	IMF3, IMF4, IMF5	IMF6, IMF7, IMF8	0.8171	2.3069	0.8278	0.7495
B	IMF0, IMF1	IMF2, IMF3, IMF4	IMF5, IMF6, IMF7, IMF8	0.8153	2.3210	0.8389	0.7302
C	IMF0, IMF1, IMF2	IMF3, IMF4	IMF5, IMF6, IMF7, IMF8	0.8656	2.1305	0.8549	0.6930

Mode A and Mode C for subsequence reconstruction, corresponding to an IMF number of 8 and 9, respectively, achieve the best prediction performance. Considering the sample entropy values of each IMF listed in Table 1, we could summarize the threshold of the sample entropy value to obtain the best reconstruction mode.

It can be inferred that subsequences whose sample entropy values are higher than 1.0 or lower than 0.2 should be reconstructed into a new re-IMF; the others are the third re-IMF, setting the principle of subsequence reconstruction in this study.

(3) Select the appropriate subsequence for secondary decomposition.

Secondary decomposition could improve model performance but consume more time; which reconstructed IMF from the subsequence reconstruction process should be re-decomposed is investigated by the comparison experiments listed in Table 5.

Table 5. Comparison of prediction models with different re-decomposition programs.

Components to be Re-Decomposed	Well-1		Well-2		Well-3		Well-4	
	R^2	Time (s)	R^2	Time (s)	R^2	Time (s)	R^2	Time (s)
re-IMF2	0.8282	865.777	0.8800	858.695	0.8472	801.331	0.8017	901.619
re-IMF1	0.8394	902.950	0.8826	867.678	0.8641	870.178	0.8334	909.153
re-IMF0	0.9235	928.145	0.9603	1001.435	0.9364	944.565	0.9483	926.474
re-IMF0, re-IMF1	0.8802	1206.589	0.9615	1267.169	0.8811	1282.281	0.9151	1205.483
re-IMF0, re-IMF1,re-IMF2	0.8546	1299.403	0.9621	1270.926	0.8781	1283.385	0.8313	1249.192

While decomposing only one subsequence, the prediction performance of the model gradually decreases from re-IMF0 to re-IMF2. On the other hand, with the increase in the number of sequences that are secondarily decomposed, the prediction accuracy of the model improves very slightly or doesn't improve; however, the computing time increases significantly. Re-IMF0 includes much complex information about the source data; further processing could capture the sufficient features to forecast more accurately. Furthermore, while low-frequency data is initially easy to forecast, additional processing leads to increased cumulative errors and calculation workload. Therefore, only the highest-frequency subsequence should be applied to the secondary decomposition procedure.

3.3. Experiment I: Comparison of Single Models

The forecasting performance of four single artificial intelligence models without data decomposition is compared in this section, including Support Vector Regression (SVR), Back Propagation (BP) Neural Network, Recurrent Neural Network (RNN) and LSTM. Based on the oil production prediction values of Well-1, the evaluation metrics are calculated and shown in Figure 4.

Evidently, LSTM has the smallest error (RMSE, MAE, MAPE) and highest accuracy (R^2) among single models. It implies that, when compared to SVR, BP and RNN, LSTM can grasp the high-sophistication features of the oil production rate more effectively, making it more suitable as a prediction method for production dynamic analysis in oilfields. However, single models still cannot satisfy the requirements of high forecast accuracy for their low-level R^2 score.

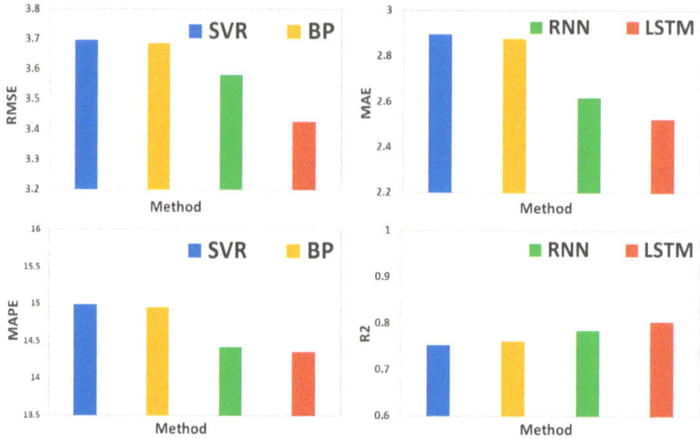

Figure 4. Evaluation results of single models.

3.4. Experiment II: Comparison of First Decomposition Methods before Sample Entropy Reconstruction

The primary objective of this scenario is to compare the performance of popular decomposition methods based on the decomposition-sample entropy reconstruction-ensemble forecasting framework. The error index values obtained by different decomposition methods, including EMD, Ensemble Empirical Mode Decomposition (EEMD) and CEEMDAN, are shown in Figure 5, while simple LSTM is the baseline.

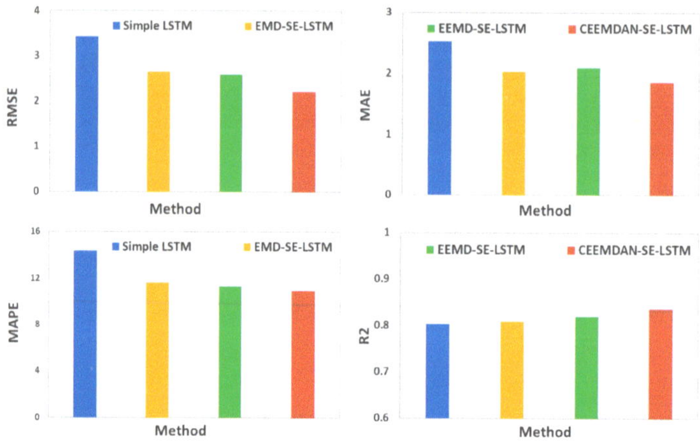

Figure 5. Performance of models with one-step decomposition compared with LSTM.

After adopting EMD, EEMD and CEEMDAN as first-stage decomposition methods, the prediction accuracy of the model has been significantly improved; even the EMD-SE-LSTM model with low accuracy (RMSE = 2.6570, MAE = 2.0306, MAPE = 11.6345, R^2 = 0.8093) is much better than the single LSTM. In contrast with EMD and EEMD, CEEMDAN is more suitable for processing well production data.

3.5. Experiment III: Comparison of Different Predictors Based on the Hybrid Structure with Primary Decomposition and Sample Entropy Reconstruction

This section investigated the performance of the hybrid structure via primary decomposition and sample entropy reconstruction with a classical predictor. SVR, BP, RNN and

LSTM were introduced as predictors in the ensemble forecasting framework, and their prediction performances are shown in Figure 6.

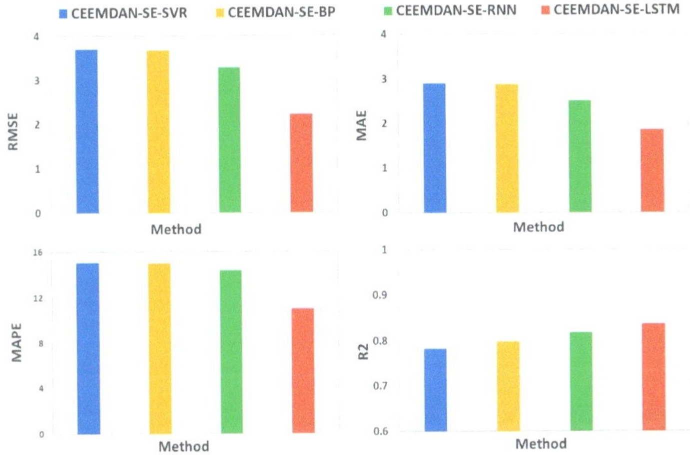

Figure 6. Evaluation indices of the hybrid forecasting structure with different predictors.

Based on the hybrid model, the performance of multiple predictors appeared to follow the following order: deep learning techniques (RNN, LSTM) > machine learning methods (SVR, BP). It could be inferred that the forecasting efficiency of LSTM was infinitely superior to traditional machine learning algorithms and RNN because of its particular structure and ability to process time series.

3.6. Experiment IV: Comparison of Different Forecasting Architectures

There are two basic structures during the time series forecast process, which are named integral architecture and corresponding architecture in this issue. Integral architecture means applying all series data to an individual model, so the input data should be a matrix. On the contrary, the corresponding structure is more complicated and accurate because it predicts all series data separately. The forecasting model's quantities are dependent on the counts of input vectors, which increases the prediction procedure's calculation time. Considering the evident advantages and disadvantages of two structures comprehensively, a hybrid architecture is established. First, it uses the corresponding structure to forecast each IMF or re-IMF, then uses the integral architecture to integrate the results of the previous step. Figure 7 exhibits the evaluation indicators for three forecasting architectures compared with a simple LSTM baseline. The calculation time of each structure is shown in Table 6.

Although the corresponding architecture has the smallest error, it requires a significant amount of calculation time. The integral architecture runs fastest but sacrifices accuracy. The hybrid structure demonstrates nearly identical performance to the corresponding architecture and saves lots of time. It combines the advantages of two basic architectures, maintaining prediction accuracy while promoting computing speed.

Table 6. Calculation time of different architectures.

Architecture	Calculation Time (s)
Simple LSTM	191.518
CEEMDAN-SE-Integral LSTM	267.187
CEEMDAN-SE-Corresponding LSTM	953.668
CEEMDAN-SE-Hybrid LSTM	554.536

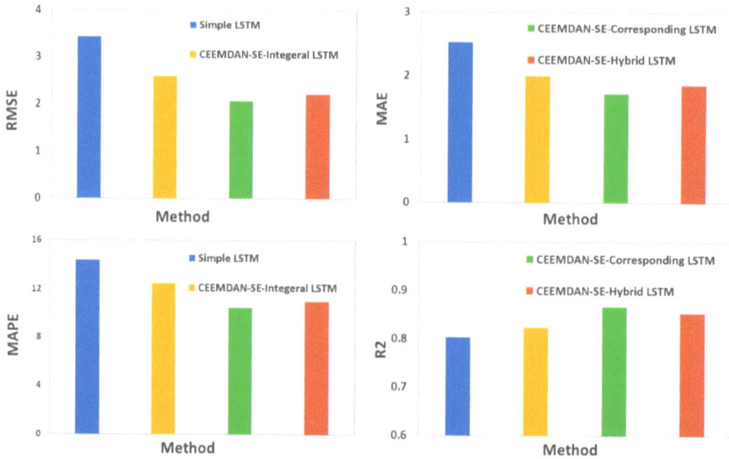

Figure 7. Performance comparison of different forecasting architectures based on LSTM.

3.7. Experiment V: Comparison of Second-Stage Decomposition Methods

After first-step processing, the raw data is decomposed into several IMFs, which consist of high-frequency sequences and low-frequency sequences; LSTM could forecast the latter more effectively. Further processing for high-frequency series data could enhance the model's performance; multi-stage decomposition is suggested. In this experiment, IMF0 or re-IMF0 is decomposed secondarily by different decomposition methods based on the hybrid forecasting architecture, including EMD, EEMD, CEEMDAN and VMD. The comparative result of these models is shown in Figure 8.

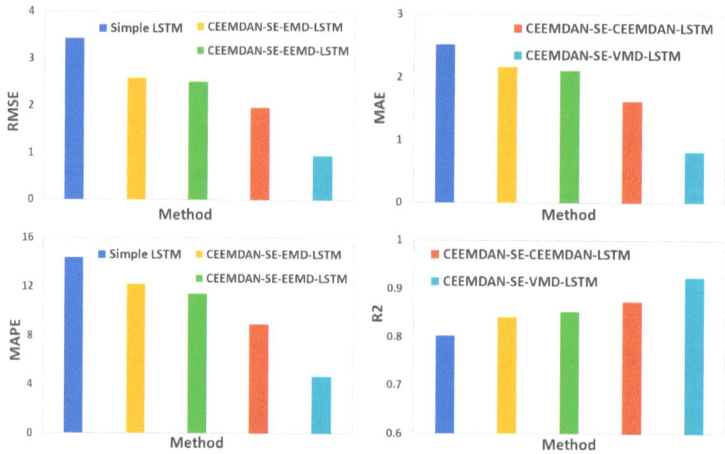

Figure 8. Results of model evaluation for the different secondary decomposition methods.

These three EMD-based methods perform better than simple LSTM; however, applying VMD to the second-stage decomposition process decreases the error evidently (RMSE decreased by 72.74%, MAE by 67.82% and MAPE by 67.66%) and promotes the R^2 score by 15.01%.

3.8. Experiment VI: Comparison of Proposed Hybrid Model with Other Forecasting Methods

To verify the proposed model's progression and creativeness, it is essential to compare it to other models that are usually utilized for time series forecasting, including BP, single

LSTM, CEEMDAN-SE-LSTM, and integral forecasting architecture based on VMD second-decomposition. The comparison with these models illustrates the value of the proposed model. The evaluation indices of these models are depicted in Figure 9.

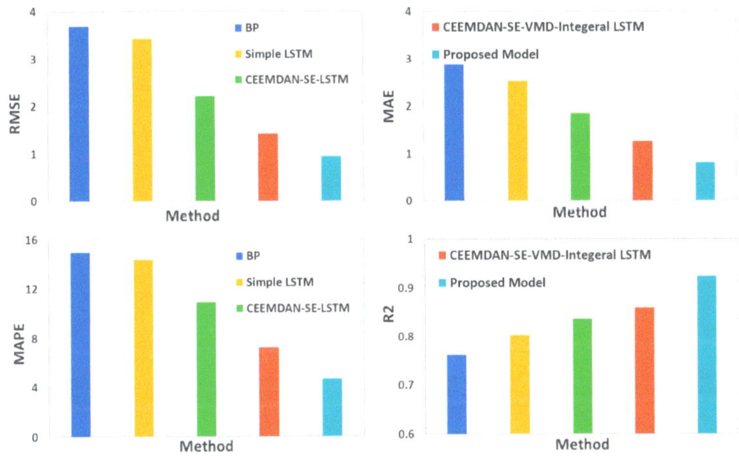

Figure 9. Comparison of the proposed model with other methods.

The proposed model has the smallest error and the highest R^2 value among the commonly used methods; the results confirmed the validity of the proposed hybrid approach for oil production forecasting. Figure 10 shows the performance of the hybrid model in predicting Well-1's oil production.

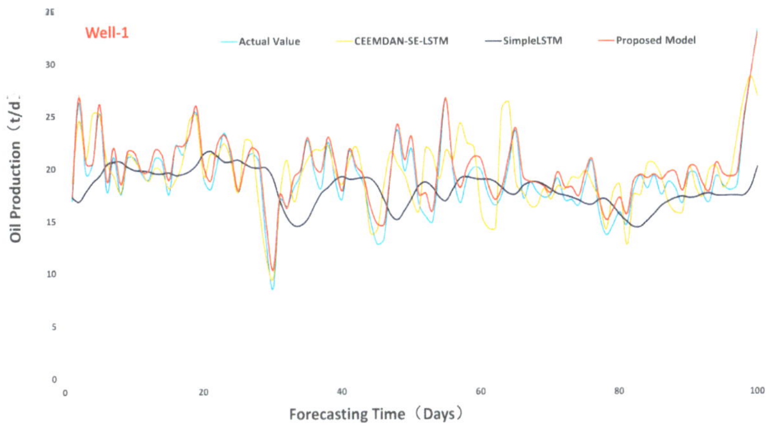

Figure 10. Forecasting performance of the proposed model for Well-1.

The hybrid structure of decomposition-reconstruction-secondary decomposition contributes to the ability to distinguish information of different frequencies and capture deeper features of oil production data, achieving more accurate prediction outcomes. In fact, frequently and abruptly changing values, as illustrated in Figures 11 and 12, remain a difficult issue in forecasting. More engineering parameters should be considered when using the model in the future.

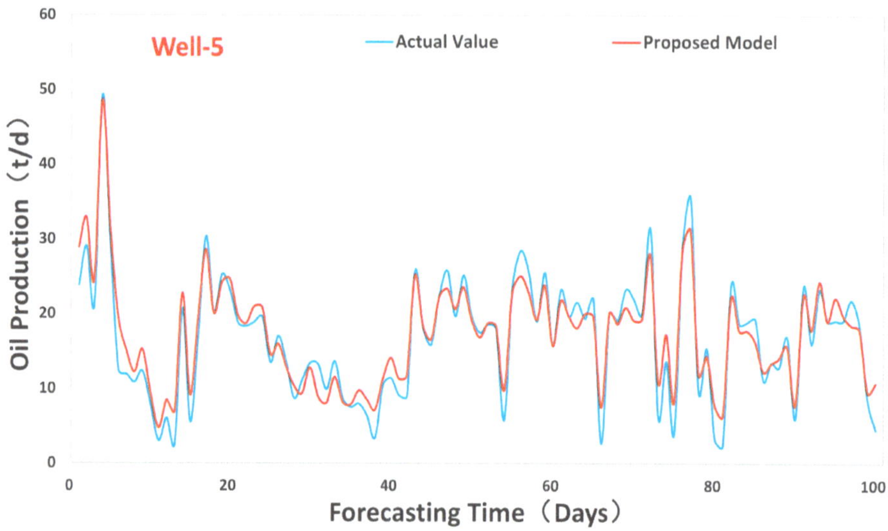

Figure 11. Forecasting performance of the proposed model for Well-5.

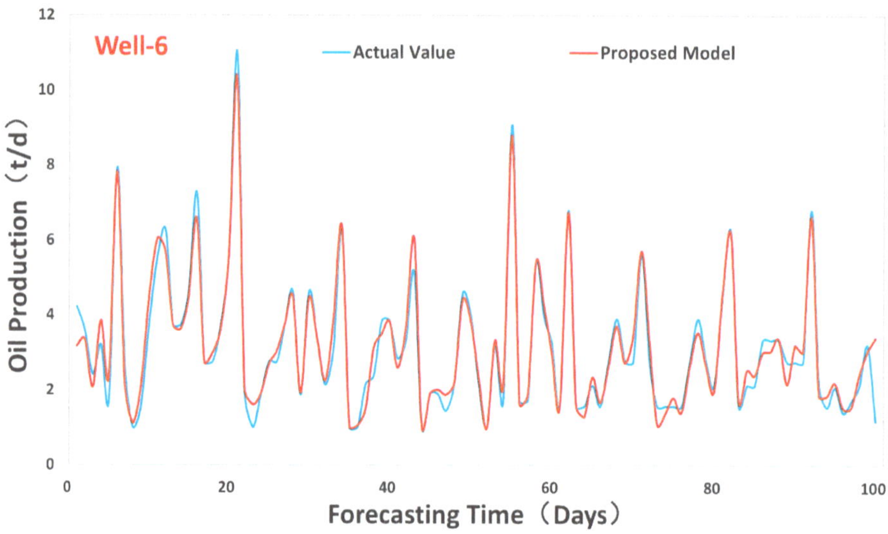

Figure 12. Forecasting performance of the proposed model for Well-6.

3.9. Experiment VII: Validations in Other Production Wells

The proposed hybrid prediction framework achieved outstanding performance in forecasting Well-1's production dynamics. We implement it for other wells' production predictions to validate the hybrid model. Figures 13–15 demonstrate the production forecasting results of three wells in the JL volcanic reservoir.

The results validate that the hybrid model proposed in this study also achieves good accuracy in forecasting other wells' production. The proposed model could obtain the characteristics and trends of production history data and make accurate predictions, providing an applicable method for reservoir production forecasting.

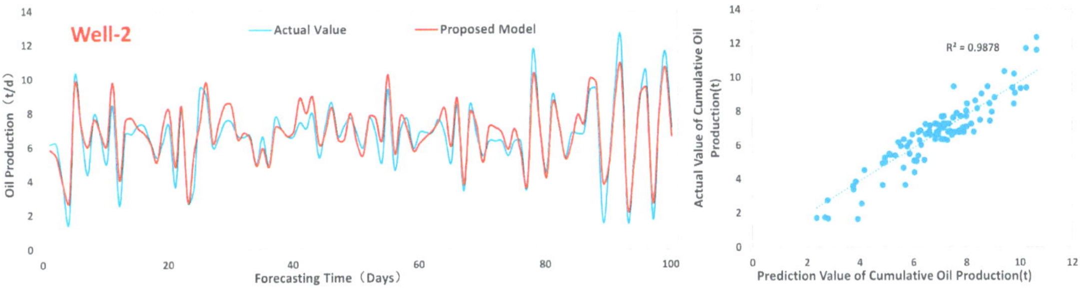

Figure 13. Production forecasting with proposed hybrid model for Well-2.

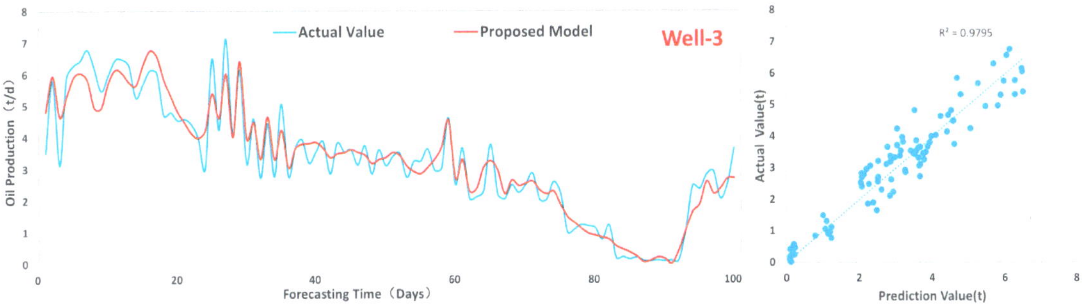

Figure 14. Forecasting result for Well-3.

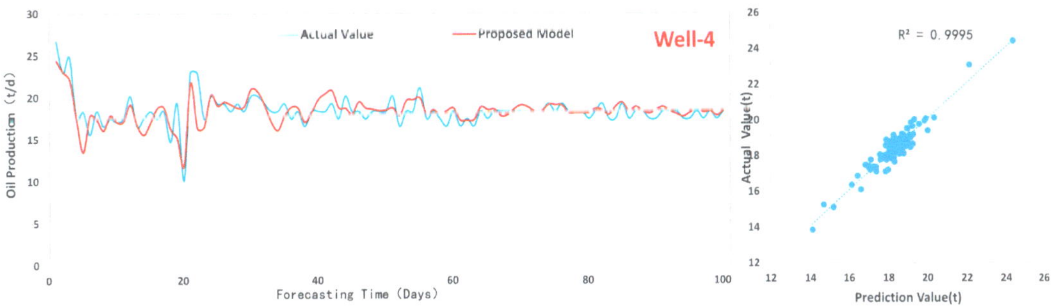

Figure 15. Prediction performance of the proposed model for Well-4.

4. Discussion

This study aims to establish a hybrid framework to analyze the production dynamics and improve the prediction accuracy of oil reservoirs. Simple models cannot capture all features of reservoir history information, making inaccurate predictions. Decomposition methods could process nonstationary, complex and low-quality data in the petroleum industry. The raw data can be transformed into subsequences, facilitating feature engineering; however, the complexity of some subsequences is still high. To enhance the model's efficiency and avoid unnecessary computing, IMFs' integration is conducted under the rules of sample entropy reconstruction determined by evaluation experiments. The remaining highest-frequency component contains the main irregular parts of the data, which inhibits the model's performance. A secondary decomposition could improve the model's prediction performance by addressing nonstationary and nonlinear issues in the highest-frequency data and fully extracting time series features. The LSTM structure, which is able to capture and store valuable features from time series data, is appropriate for the hybrid forecasting structure with high prediction accuracy. The proposed model is applied for single-well production prediction in the JL volcanic reservoir and performs outstandingly, achieving more accurate forecast results than single models and one-step decomposition structures. Moreover, the rules for reconstruction and re-decomposition make the process more automated and intelligent.

Although principles of reconstruction and secondary decomposition are specified for the proposed framework, we advise adopting the method in other reservoirs with similar geological or production characteristics. For different types of oilfields, the threshold in the rules should be adjusted according to the actual production dynamics of oil wells. Furthermore, more geological and engineering features should be considered to enhance the forecast of suddenly changing values, and the predictor in the final stage of the structure should be determined among multiple methods by comparable evaluations since there is no perfect technique for every task.

5. Conclusions

Oil production forecasting is extremely significant during oilfield development, particularly in unconventional reservoirs such as volcanic reservoirs. Traditional methods and simple machine learning algorithms cannot achieve sufficiently high accuracy in oil production forecasting results. High-frequency data from the decomposition-prediction strategy also limits the forecasting efficiency; thus, a multi-stage decomposition model is proposed in this study. The hybrid model consists of CEEMDAN, sample entropy reconstruction, secondary decomposition based on VMD and the LSTM forecasting process. Based on data-driven theory, two-stage decomposition could aid in extracting features and increasing prediction accuracy. The structure synthesizes the advantages of the integral and corresponding architecture, making it outstanding among comparative experiments. The rules of reconstruction and secondary decomposition are defined, and which subsequence should be decomposed again is also proposed, improving the workflow's generalizability. The proposed hybrid model was validated in several actual production wells, illustrating its wide applicability. We can apply the model to other reservoirs with similar geological features or similar oil production patterns, such as other volcanic reservoirs, naturally fractured reservoirs and low permeable reservoirs developed by artificial fracturing. The common property of these formations is that their oil production capacity is determined by fracture development. Besides predicting production, the method helps to understand the reservoir thoroughly and adjust the subsequent development scheme.

Author Contributions: Conceptualization, X.K.; methodology, X.K.; software, X.K.; validation, X.K.; formal analysis, X.K.; investigation, X.K.; resources, Y.L.; data curation, X.K. and D.Z.; writing—original draft preparation, X.K.; writing—review and editing, Y.L., L.X. and G.L.; visualization, X.K.; supervision, Y.L.; project administration, Y.L.; funding acquisition, Y.L. All authors have read and agreed to the published version of the manuscript.

Funding: This research was funded by National Natural Science Foundation of China (No. 51374222, No.52274048), the National Basic Research Program of China (973 Program, No. 2015CB250905), the National Major Science and Technology Projects of China (No. 2017ZX05032004-002), CNPC Major Scientific Research Project (No. 2017E-0405), SINOPEC Major Scientific Research Project (No. P18049-1), Beijing Natural Science Foundation (No. 3222037), the PetroChina Innovation Foundation (No. 2020D-5007-0203), the Science Foundation of China University of Petroleum, Beijing (No. 2462021YXZZ010) and the PetroChina perspective fundamental research project (No. 2021DJ2104).

Institutional Review Board Statement: Not applicable.

Data Availability Statement: Not applicable.

Conflicts of Interest: The authors declare no conflict of interest.

References

1. Arps, J.J. Analysis of decline curves. *Trans. AIME* **1945**, *160*, 228–247. [CrossRef]
2. Ji, J.; Yao, Y.; Huang, S.; Ma, X.; Zhang, S.; Zhang, F. Analytical model for production performance analysis of multi-fractured horizontal well in tight oil reservoirs. *J. Pet. Sci. Eng.* **2017**, *158*, 380–397. [CrossRef]
3. Sun, R.; Hu, J.; Zhang, Y.; Li, Z. A semi-analytical model for investigating the productivity of fractured horizontal wells in tight oil reservoirs with micro-fractures. *J. Pet. Sci. Eng.* **2020**, *186*, 106781. [CrossRef]
4. Alfi, M.; Hosseini, S.A. Integration of reservoir simulation, history matching, and 4D seismic for CO_2-EOR and storage at Cranfield, Mississippi, USA. *Fuel* **2016**, *175*, 116–128. [CrossRef]
5. Ning, Y.; Kazemi, H.; Tahmasebi, P. A comparative machine learning study for time series oil production forecasting: ARIMA, LSTM, and Prophet. *Comput. Geosci.* **2022**, *164*, 105126. [CrossRef]
6. Song, X.; Liu, Y.; Xue, L.; Wang, J.; Zhang, J.; Wang, J.; Jiang, L.; Cheng, Z. Time-series well performance prediction based on Long Short-Term Memory (LSTM) neural network model. *J. Pet. Sci. Eng.* **2020**, *186*, 106682. [CrossRef]
7. Wang, S.; Chen, Z.; Chen, S. Applicability of deep neural networks on production forecasting in Bakken shale reservoirs. *J. Pet. Sci. Eng.* **2019**, *179*, 112–125. [CrossRef]
8. Huang, R.; Wei, C.; Wang, B.; Yang, J.; Xu, X.; Wu, S.; Huang, S. Well performance prediction based on Long Short-Term Memory (LSTM) neural network. *J. Pet. Sci. Eng.* **2022**, *208*, 109686. [CrossRef]
9. Sagheer, A.; Kotb, M. Time series forecasting of petroleum production using deep LSTM recurrent networks. *Neurocomputing* **2019**, *323*, 203–213. [CrossRef]
10. Cheng, Y.; Yang, Y. Prediction of oil well production based on the time series model of optimized recursive neural network. *Pet. Sci. Technol.* **2021**, *39*, 303–312. [CrossRef]
11. Lu, W.; Rui, H.; Liang, C.; Jiang, L.; Zhao, S.; Li, K. A Method Based on GA-CNN-LSTM for Daily Tourist Flow Prediction at Scenic Spots. *Entropy* **2020**, *22*, 261. [CrossRef] [PubMed]
12. Yang, W.; Wang, J.; Wang, R. Research and Application of a Novel Hybrid Model Based on Data Selection and Artificial Intelligence Algorithm for Short Term Load Forecasting. *Entropy* **2017**, *19*, 52. [CrossRef]
13. Zha, W.; Liu, Y.; Wan, Y.; Luo, R.; Li, D.; Yang, S.; Xu, Y. Forecasting monthly gas field production based on the CNN-LSTM model. *Energy* **2022**, *260*, 124889. [CrossRef]
14. Fan, D.; Sun, H.; Yao, J.; Zhang, K.; Yan, X.; Sun, Z. Well production forecasting based on ARIMA-LSTM model considering manual operations. *Energy* **2021**, *220*, 119708. [CrossRef]
15. Li, W.; Wang, L.; Dong, Z.; Wang, R.; Qu, B. Reservoir production prediction with optimized artificial neural network and time series approaches. *J. Pet. Sci. Eng.* **2022**, *215*, 110586. [CrossRef]
16. Meng, F.; Xu, D.; Song, T. ATDNNS: An adaptive time–frequency decomposition neural network-based system for tropical cyclone wave height real-time forecasting. *Future Gener. Comput. Syst.* **2022**, *133*, 297–306. [CrossRef]
17. Lv, P.; Wu, Q.; Xu, J.; Shu, Y. Stock Index Prediction Based on Time Series Decomposition and Hybrid Model. *Entropy* **2022**, *24*, 146. [CrossRef]
18. Zhou, F.; Huang, Z.; Zhang, C. Carbon price forecasting based on CEEMDAN and LSTM. *Appl. Energy* **2022**, *311*, 118601. [CrossRef]
19. Liu, W.; Liu, W.D.; Gu, J. Forecasting oil production using ensemble empirical model decomposition based Long Short-Term Memory neural network. *J. Pet. Sci. Eng.* **2020**, *189*, 107013. [CrossRef]
20. Wang, F.; Zhang, D.; Min, G.; Li, J. Reservoir Production Prediction Based on Variational Mode Decomposition and Gated Recurrent Unit Networks. *IEEE Access* **2021**, *9*, 53317–53325. [CrossRef]
21. Guo, Z.H.; Zhao, W.G.; Lu, H.Y.; Wang, J.Z. Multi-step forecasting for wind speed using a modified EMD-based artificial neural network model. *Renew. Energy* **2012**, *37*, 241–249. [CrossRef]
22. Torres, M.E.; Colominas, M.A.; Schlotthauer, G.; Flandrin, P. A Complete Ensemble Empirical Mode Decomposition with Adaptive Noise. In Proceedings of the 2011 IEEE International Conference on Acoustics, Speech and Signal Processing (ICASSP), Prague, Czech Republic, 22–27 May 2011; pp. 4144–4147.
23. Dragomiretskiy, K.; Zosso, D. Variational mode decomposition. *IEEE Trans. Signal Process.* **2013**, *62*, 531–544. [CrossRef]
24. Hochreiter, S.; Schmidhuber, J. Long Short-Term Memory. *Neural Comput.* **1997**, *9*, 1735–1780. [CrossRef] [PubMed]
25. Richman, J.S.; Moorman, J.R. Physiological time-series analysis using approximate entropy and sample entropy. *Am. J. Physiol. Heart Circ. Physiol.* **2000**, *278*, H2039–H2049. [CrossRef] [PubMed]

energies

MDPI

Article

A Novel Workflow for Early Time Transient Pressure Data Interpretation in Tight Oil Reservoirs with Physical Constraints

Tongjing Liu [1], Liwu Jiang [2,*], Jinju Liu [2,*], Juan Ni [1], Xinju Liu [1] and Pengxiang Diwu [3]

1 Unconventional Petroleum Research Institute, China University of Petroleum (Beijing), Beijing 102249, China
2 Petroleum Systems Engineering, University of Regina, Regina, SK S4S 0A2, Canada
3 College of Science, China University of Petroleum (Beijing), Beijing 102249, China
* Correspondence: jiang37l@uregina.ca (L.J.); jlb670@uregina.ca (J.L.)

Abstract: In this work, a novel workflow has been proposed, validated and applied to interpret the early time transient pressure data in tight oil reservoirs with physical constraints. More specifically, the theoretical model was developed to obtain the transient pressure response for a vertical well in tight oil reservoirs with consideration of pseudo threshold pressure gradient (TPG). Then, a physical constraint between the skin factor and formation permeability has been proposed based on the physical meaning of percolation theory. This physical constraint can be applied to determine the lower limit of the skin factor which can reduce the uncertainty during the interpretation process. It is found that the influence range of the skin factor and permeability may partially overlap during the interpretation process without consideration of physical constraints. Additionally, it is found that the equivalent wellbore radius is more reasonable by considering the skin factor constraints. Furthermore, the short-time asymptotic method was applied to separate the small pressure signal at the early time period and a novel type curve was proposed to better analyze the early time pressure response. Subsequently, sensitivity analyses were conducted to investigate the influence of different parameters on the new type curves. It is found that the new type curves are more dispersed and sensitive to the parameters at the early time period which can be beneficial for the early time transient pressure analysis in a tight formation. The proposed method has been validated and then extended to a field application, demonstrating that the transient pressure for a vertical well in a tight formation can be analyzed in a reasonable and accurate manner with only early time transient pressure data.

Keywords: tight oil reservoirs; pseudo threshold pressure gradient (TPG); early time transient data; skin factor; physical constraints; new type curves

1. Introduction

Due to the declining reserves of conventional reservoirs and the growing global energy consumption, unconventional resources have received increasing reliance [1]. As tight oil reservoirs account for a significant portion of unconventional hydrocarbon resources, the efficient development of tight oil reservoirs has become a primary focus [2,3]. Pressure transient analysis is considered to be a valuable method to provide information about reservoirs by interpreting the pressure data [4–6]. For pressure transient analysis, the reservoir properties can be estimated or determined by identifying different flow regimes on the pressure derivative curve [7–10]. Since the permeability of a tight oil reservoir is extremely low, it usually takes quite a long time to achieve the infinite acting radial flow (IARF) regime and the early time transient pressure data can be difficult to interpret which makes the well testing method less attractive [11]. Therefore, the traditional pressure transient analysis method is not applicable to tight oil reservoirs and it is of fundamental and practical importance to find an appropriate method to analyze the early time transient pressure data in the tight formation in a consistent and accurate manner.

The low-velocity non-Darcy flow phenomenon can be described by the curves shown in Figure 1 where the red curve represents the traditional Darcy flow. However, at the lower pressure gradient region, fluid cannot flow through the porous media unless the pressure gradient exceeds the threshold pressure gradient (TPG). At the region where the pressure gradient larger than TPG, a linear relationship between the flow velocity and pressure gradient can be found. The existence of TPG has been observed from laboratory works [12,13] which can be explained by the effect of tight pore structures, non-Newtonian fluid, and boundary layer [14,15]. Many factors, including pore-throat size, capillary pressure, fluid saturation, and permeability can affect the value of threshold pressure gradient in a tight formation, and the reasonable range of threshold pressure gradient in a tight formation has been reported to be 0.006–0.04 MPa/m [16–18]. The TPG can not only result in nonlinear and nonhomogeneous diffusivity equations which are difficult to solve but also can affect the production performance of the tight formations [19]. Numerous attempts have been made to investigate the transient pressure behavior by considering the TPG through both numerical and analytical methods [7,20,21]. However, it is found that the TPG mainly affects the flow regime at late time periods where little information about the TPG can be found from analyzing the early time transient pressure data [7,21]. In order to better understand the reservoir properties with consideration of TPG from the transient pressure analysis method, it is urgent to seek a trustworthy method for such a purpose.

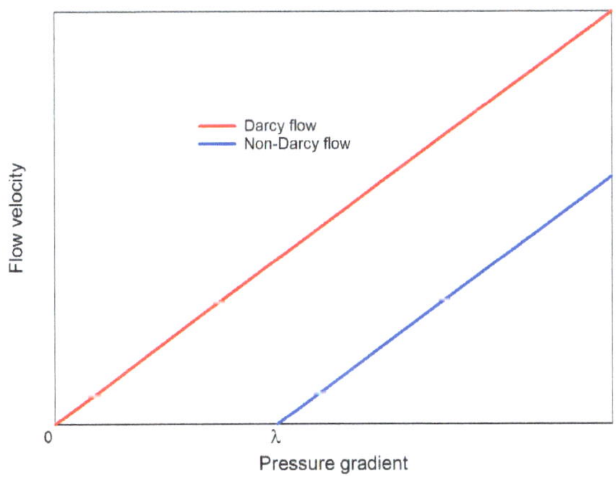

Figure 1. Schematic diagram of low velocity non-Darcy flow model.

As the lower permeability and poor flow capacity in a tight formation, it is difficult to reach the IARF regime for the well testing method where long time pressure testing operation will seriously affect the well production [11]. As such, the early time transient pressure data without enough information about the IARF regime are usually obtained during the well testing method for a tight formation [7]. For the early time transient pressure response, usually three parameters dominate the early time pressure response including the wellbore storage coefficient, skin factor, and reservoir permeability. For the latter two parameters, it is usually hard to separate their respective influence on the transient pressure curves, and there is no physical constraint for the lower limit of skin factor (usually between −6–100) [22] which can bring high ambiguity for the well testing interpretation results. Efforts have been made to analyze the early time trainset rate/pressure data by using various new type curves under various conditions for various reservoirs [23–25]. However, the TPG is usually neglected for such reservoirs which may greatly affect the transient pressure behavior for a tight formation.

In this work, a numerical model has been proposed and validated to investigate the effect of TPG, wellbore storage, and skin factor on the early time transient pressure response for a tight formation. The physical constraint for the skin factor has been proposed which can reduce the uncertainty of the interpretation results. Furthermore, a novel type curve has been developed which is capable to extract the small pressure signal at the early time period. Based on the new type curves, sensitivity analyses have been done to examine the effect of TPG, wellbore storage, and skin factor for both Darcy flow model and low-velocity non-Darcy flow model.

2. Methodology

In this work, a vertical well is located in a cylindrical tight oil reservoir with infinite boundary (see Figure 2), and the main assumptions are listed as follows,

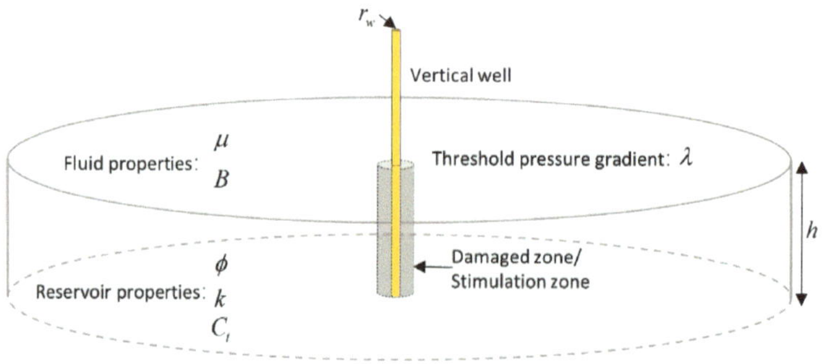

Figure 2. Schematic diagram of the reservoir and well model used in this work.

(1) The infinite reservoir is homogeneous with constant thickness;
(2) Slightly compressible single phase fluid is assumed in the formation;
(3) Fluid flow in the formation obeys the low-velocity non-Darcy flow characterized by TPG;
(4) The well production rate is constant during the production periods; and
(5) Wellbore storage and skin factor are considered, and gravity effect is ignored in this work.

2.1. Analytical Solution

The vertical well fully penetrates the formation, and the gravity effect is ignored in this work. Thus, only the radial flow towards the vertical wellbore is considered which can accurately describe the fluid flow behavior in such a reservoir, while the vertical flow can be neglected in this work. The governing equation for the isotropic reservoir with consideration of pseudo TPG can be written as follows, and the detailed derivation can be found in Appendix A,

$$\begin{cases} \frac{\partial^2 p}{\partial r^2} + \frac{1}{r}\frac{\partial p}{\partial r} - \frac{1}{r}\lambda = \frac{\phi\mu C_t}{k}\frac{\partial p}{\partial t} \\ p(r,t)|_{t=0} = p_i \\ \lim_{r\to\infty} p(r,t) = p_i \\ qB = \left[\frac{2\pi r h k}{\mu}\left(\frac{\partial p}{\partial r} - \lambda\right)\right]_{r=r_{we}} - C\frac{dp_w}{dt} \\ p_w = p(r_{we},t) \end{cases} \tag{1}$$

where p is formation pressure, r is the distance away from the wellbore, ϕ is formation porosity, μ is fluid viscosity, C_t is total formation compressibility, k is the formation permeability, t is real time, λ is the pseudo TGP, B is the oil formation volume factor, p_i is the initial formation pressure, p_w is the well bottom-hole pressure (BHP), C is the wellbore storage coefficient, S is the sin factor, and r_{we} is the equivalent wellbore radius (i.e., $r_{we} = r_w e^{-S}$).

The following dimensionless variables are defined for further analysis,

$$p_D = \frac{2\pi kh(p_i - p)}{qB\mu}, t_D = \frac{kt}{\phi\mu C_t r_w^2}, r_D = \frac{r}{r_w e^{-S}}, C_D = \frac{C}{2\pi h\phi C_t r_w^2}, \lambda_D = \frac{2\pi khr_w\lambda}{qB\mu} \tag{2}$$

Then, the dimensionless governing equations together with the boundary and initial conditions can be obtained as follows,

$$\begin{cases} \frac{\partial^2 p_D}{\partial r_D^2} + \frac{1}{r_D}\frac{\partial p_D}{\partial r_D} + \frac{\lambda_D e^{-S}}{r_D} = \frac{1}{C_D e^{2S}}\frac{\partial p_D}{\partial (t_D/C_D)} \\ p_D(r_D, t_D/C_D)\big|_{t_D/C_D=0} = 0 \\ \lim\limits_{r_D\to\infty} p_D(r_D, t_D/C_D) = 0 \\ \frac{\partial p_{wD}}{\partial (t_D/C_D)} - \frac{\partial p_D}{\partial r_D}\Big|_{r_D=1} = 1 + \lambda_D e^{-S} \\ p_{wD} = p_D(1, t_D/C_D) \end{cases} \tag{3}$$

The Laplace transform is applied with respect to t_D/C_D, which is

$$\overline{p}_D = \int_0^\infty p_D e^{-ut_D/C_D} d(t_D/C_D) \tag{4}$$

Therefore, the corresponding Laplace transform of the aforementioned governing equations can be further acquired,

$$\begin{cases} \frac{\partial^2 \overline{p}_D}{\partial r_D^2} + \frac{1}{r_D}\frac{\partial \overline{p}_D}{\partial r_D} + \frac{\lambda_D e^{-s}}{u r_D} = \frac{u}{C_D e^{2S}}\overline{p}_D \\ \overline{p_D}(r_D, u)\big|_{r_D\to\infty} = 0 \\ u\overline{p}_{wD} - \frac{\partial \overline{p}_D}{\partial r_D}\Big|_{r_D=1} = \frac{1+\lambda_D e^{-S}}{u} \\ \overline{p}_{wD} = \overline{p}_D(1, u) \end{cases} \tag{5}$$

As the governing equations are nonhomogeneous due to the existence of TPG, and the general form of the solution for the governing equations can be written in the following form on the basis of the Bessel functions,

$$\overline{p}_D(r_D) = A \cdot I_0(\beta r_D) + BK_0(\beta r_D) + \frac{M}{\beta}K_0(\beta r_D)\int_\beta^{\beta r_D} I_0(\xi)d\xi + \frac{M}{\beta}I_0(\beta r_D)\int_{\beta r_D}^\infty K_0(\xi)d\xi \tag{6}$$

where $M = \lambda_D e^{-S}$, $\beta = (u/C_D e^{2S})^{0.5}$, $I_0(x)$, and $K_0(x)$ are the zero-order modified Bessel function of the first and second kind, respectively.

By applying the boundary conditions, the constant A and B can be inversely obtained. Then, the analytical pressure solution in the Laplace domain can be written as follows,

$$\begin{aligned} \overline{p}_{wD} = {} & \frac{K_0(\beta)}{uK_0(\beta)+\beta K_1(\beta)}\left[\frac{1+\lambda_D e^{-S}}{u} - u\frac{M}{\beta}I_0(\beta)\int_\beta^\infty K_0(\xi)d\xi + MI_1(\beta)\int_\beta^\infty K_0(\xi)d\xi\right] \\ & + \frac{M}{\beta}I_0(\beta)\int_\beta^\infty K_0(\xi)d\xi \end{aligned} \tag{7}$$

where $I_1(x)$, and $K_1(x)$ are the first-order modified Bessel function of the first and second kind, respectively. After obtaining the pressure solution in the Laplace domain, the well bottom hole pressure solution in the real-time domain p_{wD} can be inversely calculated by the Stehfest inverse algorithm [26]. In this work, to obtain more precise pressure solutions for the early time periods, the Stehfest number is chosen to be 12.

2.2. Skin Factor Constraint

Skin factor is used to characterize the near wellbore conditions and the connectivity between the well and the formation. The positive and negative skin factors represent the damaged and stimulated near wellbore conditions. Usually, the range of skin factor can be treated in the range of −6~+100 in the practical case [17]. However, no theoretical proof has been made to define the lower limit of this factor, and the very low skin factor can produce unrealistic or unphysical phenomenon even though the transient pressure data has been well matched. In this work, the lower limit of the skin factor can be theoretically obtained which can apparently reduce the ambiguity during the pressure data matching process, especially with a relatively short testing duration.

The definition of the skin factor is [22],

$$\Delta p_{skin} = [p(r_w,t)_{ideal} - p(r_w,t)_{actual}] = S\frac{q\mu}{2\pi kh} \tag{8}$$

where $p(r_w,t)_{ideal}$ is the BHP for the ideal case, and $p(r_w,t)_{actual}$ is the BHP for the actual case.

For the ideal case without skin factor, the analytical solution can be written as follows,

$$p(r_w,t)_{ideal} = p_i - \frac{q\mu}{4\pi kh}\left[-Ei\left(-\frac{r_w^2}{4\eta t}\right)\right] \tag{9}$$

where η is the pressure diffusivity coefficient (i.e., $\eta = k/(\phi\mu C_t)$). Substitute Equation (9) into Equation (8), we can obtain,

$$p_i - p(r_w,t)_{actual} = \frac{q\mu}{4\pi kh}\left[-Ei\left(-\frac{r_w^2}{4\eta t}\right)\right] + S\frac{q\mu}{2\pi kh} \tag{10}$$

For the practical case, the BHP cannot be higher than the initial pressure after production, and the following physical constraint can be obtained,

$$S > \frac{1}{2}Ei\left(-\frac{r_w^2}{4\eta t}\right) \tag{11}$$

The exponential integral function $Ei(-x)$ is inconvenient for practical use, when $0 < x \le 0.01$, $Ei(-x)$ can be approximated as the following form with high accuracy (see Figure 3).

$$Ei(-x) \approx -\ln\frac{e^{-0.5772}}{x}, \quad 0 < x \le 0.01 \tag{12}$$

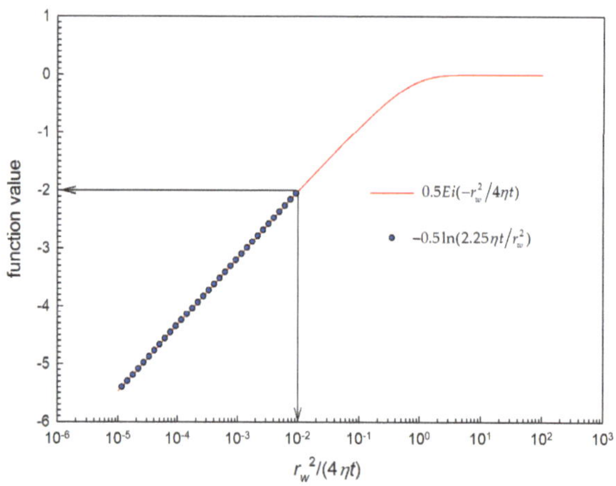

Figure 3. Different function curves representing the physical constraint of skin factor.

Additionally, as the function $Ei(-x)$ is negative, and the larger the absolute value of x the smaller the value of $Ei(-x)$. Thus, the following constraint of skin factor can be obtained,

$$S\begin{cases} > \frac{1}{2}Ei(-0.01) & \frac{r_w^2}{4\eta t} \ge 0.01 \\ > -\frac{1}{2}\ln\left(\frac{2.25\eta t}{r_w^2}\right) & \frac{r_w^2}{4\eta t} < 0.01 \end{cases} \tag{13}$$

In the field well-testing application, the minimum test time is chosen to be 10 s which is smaller enough for the transient pressure test data obtained. As the time becomes larger, the smaller value of $r_w^2/(4\eta t)$ can result in a smaller value of the lower limit of skin factor. To maintain the physical constraint for the whole testing period, the lower limit for the skin factor can be further obtained as follows,

$$S\begin{cases} > -2 & \frac{r_w^2}{4\eta} \geq 0.1 \\ > -\frac{1}{2}\ln\left(\frac{22.5\eta}{r_w^2}\right) & \frac{r_w^2}{4\eta} < 0.1 \end{cases} \tag{14}$$

Now, the constraints for the skin factor have been theoretically obtained which can be convenient for the practical application during the well-testing interpretation process.

2.3. Applicability Analysis of G-B Type Curves

Gringarten et al. [27] improved the Agarwal-Ramey type curves [28] and obtained the Gringarten type curves. Then, Bourdet et al. [29] creatively proposed the Bourdet pressure derivative (i.e., $p'_{wD} = dp_{wD}/dln(t_D/C_D)$) curves, and later the two type curves were combined into Gringarten-Bourdet type curves. The G-B type curves can describe the transient pressure behavior for a vertical well with a constant wellbore storage coefficient and skin factor in a reservoir with an infinite outer boundary. The dimensionless pressure and the pressure derivative functions are both plotted as a function of dimensionless time t_D/C_D in a single chart (see Figure 4). This chart is easy to distinguish reservoir types and flow stages, making it easier to obtain a unique fitting curve.

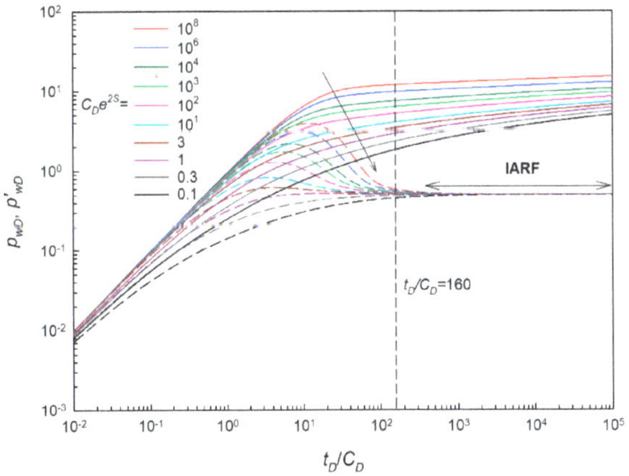

Figure 4. G-B type curves with different values of C_De^{2S}.

For the transient pressure analysis, the main flow regime of interest is the infinite acting radial flow (IARF) which can be easily identified from the log-log plot (i.e., horizontal line with Bourdet derivative equals 0.5). Additionally, combined the early time wellbore storage regime (i.e., unit slope line) with the IARF regime, the fitting or match of the pressure response would be more immediate and unique. However, there are several limitations during the well test interpretation process for tight formations. For example, in a tight formation with permeability equals 0.5 mD, reservoir thickness of 10 m, viscosity of 0.5 mPa·s, and wellbore storage coefficient of 0.5 m^3/MPa, the dimension time t_D/C_D equals 160 when the testing time equals 15 days. As shown in the following G-B type curves, the IARF regime cannot be reached even though the testing time has lasted for 15 days. If the IARF regime shall be shown on the log-log plot, the test time would be up to 100 days, which would obviously and seriously impact the well production and make the well testing method unattractive for the tight formations. Therefore, it is of fundamental and practical significance to find a method to interpret the early time transient pressure response without a possible IARF regime.

2.4. Short-Time Asymptotic Solution

For the early time transient pressure data, it is very difficult for us to identify the wellbore storage and skin factor as the pressure and its derivative curves usually overlap in a unit-slope line for the traditional G-B type curves. Recall that the pressure derivative proposed by Bourdet et al. [29] the essence of this term is actually a mathematical transformation that can separate the small pressure variance and then amplify this small signal through the product of time. The larger the time, the greater the amplification factor becomes. However, this transformation is not valid for the early time period as the pressure derivative approximately equals t_D/C_D resulting in a unit-slope line. Therefore, for the pressure response data without radial flow regimes, strong no-uniqueness can be found during the transient pressure interpretation process.

Based on the idea of Bourdet's pressure derivative, we want to extract the influence factor at the early times which is more sensitive on the C_D, S, and λ_D by mathematic transformation. The short-time asymptotic method can be applied to obtain this objective. The early time pressure solution is expanded and only the high-order series of the solution is separated which represents the small signal during the early time. In this way, more accurate type curves can be obtained which may be more sensitive on C_D, S, and λ_D during the early times and can obviously reduce the non-uniqueness of the interpretation results.

When the time is relatively smaller in the real-time domain, the Laplace variable u and β can be very large in the Laplace domain. Thus, the $K_0(x)$ and $K_1(x)$ can be asymptotically expanded as follows when $\beta \to 0$ [30],

$$K_0(\beta) = \sqrt{\frac{\pi}{2\beta}}e^{-\beta}\left[1 - \frac{1}{8\beta} + O(\beta^{-1})\right], \quad K_1(\beta) = \sqrt{\frac{\pi}{2\beta}}e^{-\beta}\left[1 + \frac{3}{8\beta} + O(\beta^{-1})\right] \tag{15}$$

Substitute Equation (14) into the solution in Equation (7), we have,

$$\overline{p}_{wD} = \left(1 + \lambda_D e^{-S}\right)\left[u^{-2} - \frac{1}{\sqrt{C_D e^{2S}}}u^{-5/2} + O\left(u^{-5/2}\right)\right] \tag{16}$$

Neglecting the high-order term with small value, then transforming the solution to the real-time domain with respect to t_D/C_D yields,

$$p_{wD} = \left(1 + \lambda_D e^{-S}\right)\left[\frac{t_D}{C_D} - \frac{4}{3\sqrt{\pi C_D e^{2S}}}\left(\frac{t_D}{C_D}\right)^{3/2}\right] \tag{17}$$

Based on the idea of parameters proposed in the literature [30], a novel parameter was defined to amplify the small variable for the early time solution in Equation (17) in this work,

$$\omega = \left|\frac{3}{2}\frac{p_{wD}}{t_D/C_D} - \frac{1}{2}\frac{dp_{wD}}{d(t_D/C_D)} - 1\right| = \left|\lambda_D e^{-S} - \frac{\left(1 + \lambda_D e^{-S}\right)}{\sqrt{\pi C_D e^{2S}}}\left(\frac{t_D}{C_D}\right)^{1/2}\right| \tag{18}$$

It can be found from the above equation that only the contribution of C_D, S, and λ_D are considered, and these constructed curves are more sensitive to C_D, S, and λ_D other than a unit slope line with little information in the G-B type curves for the early times. Additionally, this curve has a horizontal asymptote (i.e., $\lambda_D e^{-S}$) as the time t_D/C_D approaches zero when the TPG is considered. If TPG is not considered, this curve approaches the line $(t_D/C_D)^{0.5}/(\pi C_D e^{2S})^{0.5}$ when the time is small.

3. Model Validation

To verify the solutions obtained in this work, the transient pressure behavior was compared with those from previous work [31]. The low-velocity non-Darcy flow model was considered for the tight formation with the following main parameter, C_De^{2S} = 1, λ_De^{-S} = 0.0001, 0.01. As shown in Figure 3, there exists a good agreement between the results from this work and those from the literature confirming the accuracy of the solutions in this work to model the transient pressure behavior in a tight formation with consideration of the pseudo TPG. Additionally, it can be found from Figure 5, the existence of TPG can result in an increase of pressure and its derivative curves at the late time period. Additionally, the traditional IARF regime with a horizontal line at late times cannot be observed which can cause more uncertainty during the well-testing interpretation process as the most important flow regime characteristic (i.e., horizontal line with Bourdet derivative equals 0.5) has been changed due to the existence of TPG.

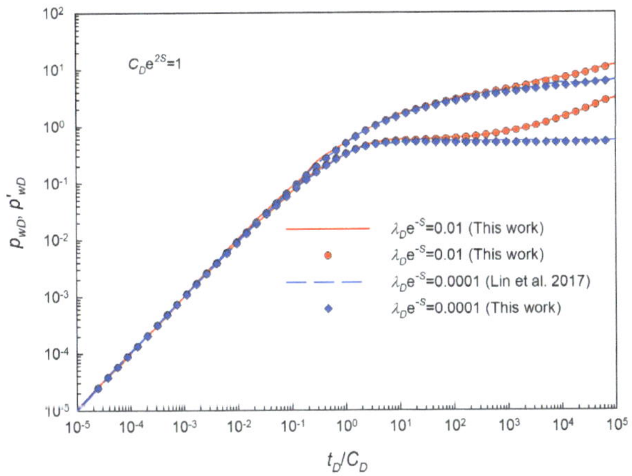

Figure 5. Comparison of transient pressure response obtained from this work and those from literature [26] for a vertical well in a tight formation with consideration of TPG.

4. Results and Discussion

In this section, the rationality of the skin factor constraints has been clarified, and the novel type curves suitable for the early time transient pressure analysis have been generated. The sensitivity analysis has been conducted to investigate the effects of some main parameters on the new type curves.

4.1. Reasonability Analysis for the Physical Constraint of Skin Factor

During the well testing interpretation for the tight oil reservoirs, the constraint conditions of skin factor are usually not considered, which would lead to doubts about the interpretation results. Specifically, the skin factor reflects the pollution or improvement in the vicinity of the wellbore, and the permeability near the wellbore region represents the seepage capacity of the formation. The conceptual scope of these two should be different.

Table 1 shows the interpretation results of a typical tight oil reservoir in the eastern part of China (the case where the skin factor is obviously abnormal and less than −5 has been removed). The current interpretation results have proved that the (inner zone) permeability can be greatly affected by the value of the skin factor. It can be found from the following table that the skin factor is between −4.89 and −2.60 when the skin factor constraint conditions are not considered, resulting in the corresponding equivalent well diameter between 0.8~8.0 m, and the average permeability is 3.92 mD. When considering the skin factor constraints, the skin factor is between −0.75~0.28, and the corresponding equivalent well diameter is between 0.05~0.13 m, and the average permeability is 16.45 mD.

Table 1. Interpretation results for the transient pressure with and without skin factor constraints.

Well Name	Without Constraints Skin Factor	Without Constraints Permeability/mD	With Constraints Skin Factor	With Constraints Permeability/mD
Well1	−2.88	4.22	−0.45	10.67
Well2	−3.79	5.23	−0.50	25.06
Well3	−3.82	1.20	−0.11	8.59
Well4	−4.89	1.90	−0.50	35.60
Well5	−3.74	1.78	−0.60	7.08
Well6	−3.12	6.93	−0.50	20.00
Well7	−3.02	4.48	−0.50	29.40
Well8	−2.82	6.45	−0.50	23.10
Well9	−3.64	3.59	−0.50	10.28
Well10	−3.22	3.06	−0.01	8.16
Well11	−3.32	15.73	−0.75	62.75
Well12	−3.65	10.11	−0.25	72.00
Well13	−3.60	8.99	−0.75	24.75
Well14	−3.00	1.21	−0.36	5.62
Well15	−4.50	0.95	−0.50	2.57
Well16	−4.66	1.05	−0.50	6.93
Well17	−3.75	8.51	−0.50	9.95
Well18	−2.60	0.40	−0.50	0.37
Well19	−3.61	0.26	0.28	0.84
Well20	−2.71	1.55	0.12	1.67
Well21	−4.18	2.35	−0.40	11.70
Well22	−2.75	0.22	−0.50	0.63
Well23	−3.68	0.08	−0.50	0.65

From a numerical point of view, the equivalent well diameter is more reasonable after considering the constraint conditions of the skin factor. Moreover, as the value of the skin factor has been confined to guarantee the correctness of the pressure solution, the influence of the skin factor can be greatly excluded for the early time transient pressure response which brings little interference to the interpretation of the permeability for the near wellbore region. As shown in Figure 6, the permeability obtained after considering the constraints of the skin factor is about 4 times that of the original result without constraints, which can reflect a more realistic seepage capacity of the formation near the wellbore which may be artificially fractured.

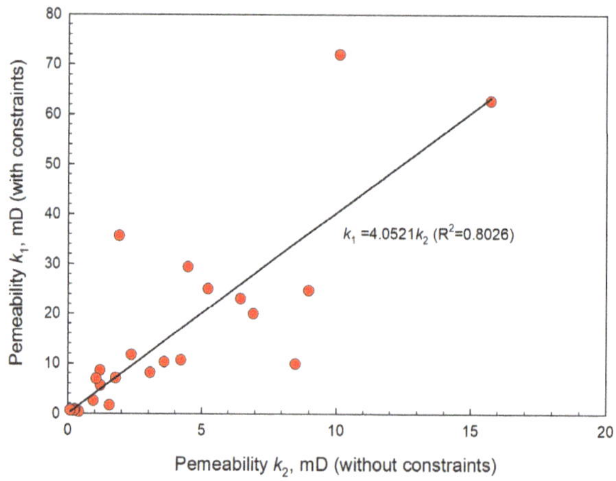

Figure 6. Effect of the skin factor constraint on the formation permeability near the wellbore interpreted from well-testing results.

4.2. Sensitivity Analysis Based on New Type Curves

4.2.1. New Type Curves for Darcy Flow Model

The low-velocity non-Darcy flow is not considered and the new type curves for the Darcy flow model are generated (i.e., $\lambda_D = 0$). The main parameters affecting the type curves are the combination of C_D and S which are the same as the G-B type curves. However, a new curve reflecting the relationship between the newly defined parameter ω and the dimensionless time t_D/C_D is added to the traditional G-B type curves. Figure 7 shows the influence of parameter C_De^{2S} on the new type curves. It can be found at the early time wellbore storage (WBS) regime, all the pressure and its derivative curves are coinciding lines with unit-slope, little information can be obtained from the transient pressure response during this period only by analyzing the pressure and pressure derivative curves. The following flow regime is the transient flow regime which can be affected by the skin factor and the permeability. The larger the value of C_De^{2S}, the higher the position of the pressure and its derivative curves. However, the differences in the pressure and its derivative curves during the transition flow regime between different cases are not large enough which can bring high uncertainty for the well testing analysis and interpretation. The last flow regime is the IARF regime, all the pressure derivative curves overlap with a horizontal line for different cases. However, such flow regime is difficult to appear for the well testing in a tight formation where the low permeability can greatly delay its arrival.

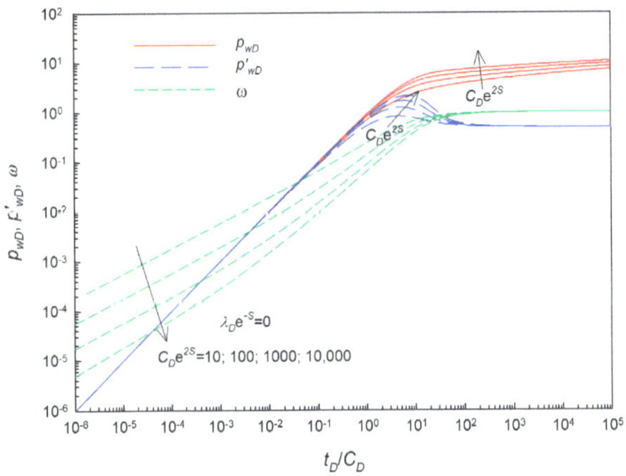

Figure 7. Effect of C_De^{2S} on the new type curves for Darcy flow model without TPG.

The green lines in Figure 7 represent the relationship between the newly defined parameter ω and the dimensionless time t_D/C_D. It can be found that these ω curves can be highly dispersed at the early time period which can be beneficial to the analysis and interpretation of the early time pressure response. The ω curves are sensitive for the parameter C_De^{2S} at the early time period, and such new curves can be used for the early time transient pressure analysis. Additionally, it can be found that all the ω curves coincide into a horizontal line with its value equals unity. Even though it can be treated as a characteristic line similar to the horizontal derivative line in IARF regime, it is not very useful for the analysis of the early time pressure response where such a horizontal line may be not present within a limited testing time. Moreover, it can be found that the ω curve shows a straight line and its slope is affected by the value of C_De^{2S} which has been discussed early after the definition of this parameter.

4.2.2. Effect of C_De^{2S} on New Type Curves for Non-Darcy Flow Model

For the low-velocity non-Darcy flow model, the TPG is considered (i.e., $\lambda_D \neq 0$). The main influencing factors on the type curves include the following combinations: C_De^{2S} and

$\lambda_D e^{-S}$. Firstly, new type curves are generated with different values of $C_D e^{2S}$ while keeping $\lambda_D e^{-S}$ constant. Figure 8 displays the new type curves with various values of $C_D e^{2S}$. It can be found that the pressure and its derivative curves are more dispersed at the transient flow regime and IARF regime due to the existence of TPG. Additionally, the larger the value of $C_D e^{2S}$, the higher the value of pressure and its derivative. As for the ω curves shown in this figure, it is more complicated than those for the Darcy flow model in Figure 7. At the starting time, as the time approaches zero, the ω curves stabilized near a horizontal line whose value equals $\lambda_D e^{-S}$. As explained early, as time becomes smaller enough, the new ω curves have a horizontal asymptote (i.e., $\lambda_D e^{-S}$). As time goes on, the value of the second term in Equation (18) becomes larger which can be in the same order of magnitude as the first term $\lambda_D e^{-S}$. Thus, a singularity point occurs at this curve where the value of the first term equals the second term. After that, the influence of the first term in Equation (18) can get weaker and weaker, and the ω curves gradually approach a straight line. At the late IARF regime, all the ω curves coincide with each other. Moreover, it can be found from Figure 8, the larger the value of $C_D e^{2S}$, the later the appearance of the singularity point of the ω curves, and the smaller the value of the ω function after the singularity point. These specific features can be easily explained by the definition of the ω function where the parameter $C_D e^{2S}$ is in the denominator position of the second term of Equation (18).

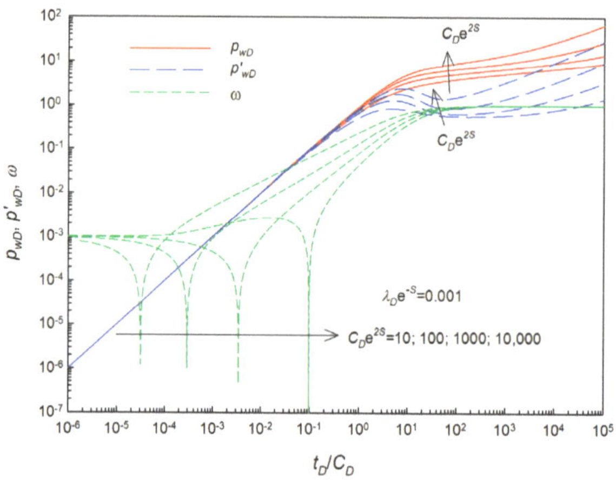

Figure 8. Effect of $C_D e^{2S}$ on the new type curves for non-Darcy flow model.

4.2.3. Effect of $\lambda_D e^{-S}$ on New Type Curves for Non-Darcy Flow Model

Another important combination parameter affecting the transient pressure type curves for the non-Darcy flow model is $\lambda_D e^{-S}$ where λ_D is the dimensionless TPG. Parameter $C_D e^{2S}$ has been assumed as constant (i.e., $C_D e^{2S}$ = 100) during the generation of the following type curves. Figure 9 presents the influence of $\lambda_D e^{-S}$ on the new type curves. As reported in previous work [7], the existence of λ_D brings extra resistance to the fluid flow in the tight formation and results in a higher value of pressure and pressure derivative curves. The larger the value of $\lambda_D e^{-S}$, the higher the position of the pressure and its derivative curves in the type curves. For the newly defined ω curves shown in Figure 9, the changing trend of such curves is similar to that discussed previously. There is a horizontal line at the beginning, then a concave shape with a singular point, followed by an approximately straight line, and finally a horizontal line with a value of 1. It can be found from the figure that, the larger the value of $\lambda_D e^{-S}$, the higher the position of the horizontal line at the beginning and the later appearance of the singularity point. These features can also be readily explained by the definition of parameter ω. At the beginning, the time approaches zero and the ω function has a horizontal asymptote with a value of $\lambda_D e^{-S}$. Additionally, it will take a larger time to make the value of the second term in Equation (18) almost equal to the value of the first term in the same equation. Moreover, the newly developed type curves are more dispersive and more sensitive to parameters at the early time period, which is very useful for us to analyze the early time pressure data with limited testing time.

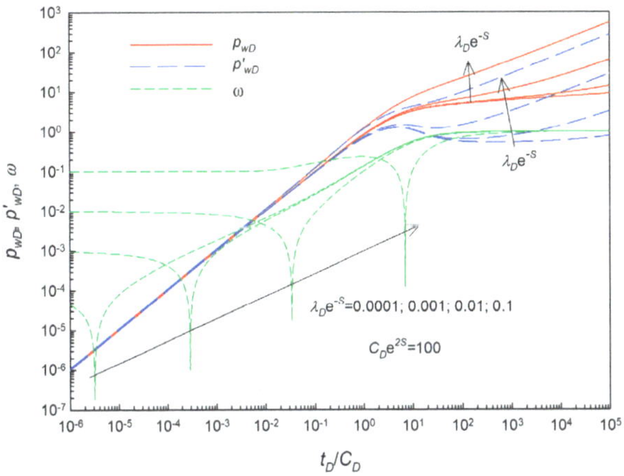

Figure 9. Effect of $\lambda_D e^{-S}$ on the new type curves for non-Darcy flow model.

4.3. Discussion

Therefore, the novel workflow for the early time transient pressure analysis can be summarized as follows: firstly generate the physical constraint for the skin factor which can greatly reduce the uncertainty for the following analysis; secondly generate the new type curves including the pressure curves, derivative curves and ω curves on a log-log plot; then generating the curves using the field well-testing data and compare with the newly generated type curves; finally the wellbore storage coefficient, skin factor, formation permeability, and threshold pressure gradient can be accurately obtained with low ambiguity.

The advantage of this work can be summarized as follows, through such workflow mentioned above, the ambiguity during the interpretation of the early time transient pressure data can be greatly reduced, and the final interpretation results can be more reliable and accurate which can reflect the realistic underground wellbore and reservoir properties. However, as the high-order term of the pressure solution shall be used to generate the new type curves which need very precise pressure data at the early times. Thus, the requirements for our well testing process and device will be more rigorous, but this is not a big challenge for the pressure gauge with high accuracy and resolution nowadays.

5. Conclusions

(1) The physical constraint of skin factor has been analyzed and the lower limit of skin factor has been obtained for practical use. The influence range of the skin factor and permeability may partially overlap during early time period without consideration of physical constraints. By considering the skin factor constraints, the interpretation parameters including the equivalent wellbore radius, and permeability near the wellbore region are more accurate and reliable.

(2) The traditional G-B type curves fail to analyze the early time transient pressure data without enough information about the IARF regime, and a novel type curve for analyzing the early time transient pressure test in a tight formation has been proposed. The novel proposed type curves can extract the small pressure signal during the early time period which are more dispersed and more sensitive for the parameters including λ_D, C_D, and S.

(3) The new ω curves show a horizontal asymptote with a value of $\lambda_D e^{-S}$, then a concave shape with a singular point, followed by an approximately straight line, and finally a horizontal line with value of 1.
(4) The larger the value of $C_D e^{2S}$ and $\lambda_D e^{-S}$, the later appearance of the singularity point for the ω curves; and the larger the value of $\lambda_D e^{-S}$, the higher the position of the horizontal asymptote at the beginning.
(5) A novel workflow has been proposed with the following features, the skin factor constraint can reduce the ambiguity and increase the rationality of interpretation results. The novel type curves are more beneficial to the analysis of the early time well testing data which are more suitable for the early time transient pressure interpretation in a tight formation.

Author Contributions: Conceptualization, T.L. and J.L.; methodology, T.L. and L.J.; software, P.D.; validation, J.N. and X.L.; formal analysis, J.N. and X.L.; writing—original draft preparation, T.L and J.L.; writing—review and editing, L.J; funding acquisition, T.L. All authors have read and agreed to the published version of the manuscript.

Funding: This research was funded by National Key Research and Development Program of China, grant number 2018YFB0605500.

Institutional Review Board Statement: Not applicable.

Informed Consent Statement: Not applicable.

Data Availability Statement: Not applicable.

Conflicts of Interest: The authors declare no conflict of interest.

Appendix A

According to the assumptions made in Section 2, the fluid flow equation considering the TPG λ can be described as follows.

$$v = \begin{cases} -\frac{k}{\mu}(\nabla p - \lambda) |\nabla p| \geq \lambda \\ 0 |\nabla p| < \lambda \end{cases} \tag{A1}$$

Additionally, the fluid is slightly compressible, based on the definition of fluid compressibility and rock compressibility, the following equations can be obtained to describe the fluid density/porosity change with respect to pressure change,

$$\rho = \rho_o[1 + C_l(p - p_o)] \tag{A2}$$

$$\phi = \phi_o[1 + C_r(p - p_o)] \tag{A3}$$

where C_l and C_r are the fluid and rock compressibility, respectively.

For the radial geometry reservoir, the following material balance equation can be obtained for a thin ring from r to $r + \Delta r$,

$$-\Delta t 2\pi(r + \Delta r) h v \rho|_{r+\Delta r} + \Delta t 2\pi r h v \rho|_r = \left[\pi(r + \Delta r)^2 - \pi r^2\right]\left[h\phi\rho|_{t+\Delta t} - h\phi\rho|_t\right] \tag{A4}$$

Dividing by the term $\Delta r \cdot \Delta t$, and let $\Delta r \to 0$, and $\Delta t \to 0$, the following diffusivity equation can be obtained,

$$\frac{1}{r} \cdot \frac{\partial(r\rho v)}{\partial r} = -\frac{\partial(\rho\phi)}{\partial t} \tag{A5}$$

Substitute Equation (A1) into Equation (A5) and neglect the first derivative squared term, the governing equation for the fluid flow in the tight formation considering TPG can be written as follows,

$$\frac{\partial^2 p}{\partial r^2} + \frac{1}{r}\frac{\partial p}{\partial r} - \frac{1}{r}\lambda = \frac{\phi\mu C_t}{k}\frac{\partial p}{\partial t} \tag{A6}$$

where the total compressibility $C_t = C_l + C_r$.

When the skin factor is considered, the inner boundary condition at the equivalent wellbore radius (i.e., $r_{we} = r_w e^{-S}$) can be written as,

$$p_w = p(r_{we}, t) \tag{A7}$$

Additionally, when the wellbore storage effect is considered, the well production consists of two parts, including the fluid provided by the wellbore storage effect and the fluid flow from the reservoir to the wellbore,

$$qB = \left[\frac{2\pi rhk}{\mu}(\frac{\partial p}{\partial r} - \lambda)\right]_{r=r_{we}} - C\frac{dp_w}{dt} \tag{A8}$$

The initial condition and outer boundary condition can be easily described as follows,

$$p(r,t)|_{t=0} = p_i \tag{A9}$$

$$\lim_{r\to\infty} p(r,t) = p_i \tag{A10}$$

References

1. Mohammed, S.; Anumah, P.; Sarkodie-Kyeremeh, J.; Acheaw, E. A production-based model for a fractured well in unconventional reservoirs. *J. Pet. Sci. Eng.* **2021**, *206*, 109036. [CrossRef]
2. Li, H.; Guo, H.; Yang, Z.; Wang, X.; Sun, Y.; Xu, H.; Zhang, H.; Lu, H.; Meng, H. Boundary retention layer influence on permeability of tight reservoir. *J. Pet. Sci. Eng.* **2018**, *168*, 562–568. [CrossRef]
3. Wu, Z.; Cui, C.; Jia, P.; Wang, Z.; Sui, Y. Advances and challenges in hydraulic fracturing of tight reservoirs: A critical review. *Energy Geosci.* **2022**, *3*, 427–435. [CrossRef]
4. Chu, H.; Liao, X.; Chen, Z.; Zhao, X.; Liu, W.; Zou, J. Pressure transient analysis in fractured reservoirs with poorly connected fractures. *J. Nat. Gas Sci. Eng.* **2019**, *67*, 30–42. [CrossRef]
5. Mohammed, I.; Olayiwola, T.O.; Alkathim, M.; Awotunde, A.A.; Alafnan, S.F. A review of pressure transient analysis in reservoirs with natural fractures, vugs and/or caves. *Pet. Sci.* **2020**, *18*, 154–172. [CrossRef]
6. Jiang, L.; Liu, J.; Liu, T.; Yang, D. Production decline analysis for a fractured vertical well with reorientated fractures in an anisotropic formation with an arbitrary shape using the boundary element method. *J. Pet. Sci. Eng.* **2022**, *208*, 109213. [CrossRef]
7. Diwu, P.; Liu, T.; You, Z.; Jiang, B.; Zhou, J. Effect of low velocity non-Darcy flow on pressure response in shale and tight oil reservoirs. *Fuel* **2018**, *216*, 398–406. [CrossRef]
8. Dejam, M.; Hassanzadeh, H.; Chen, Z. Semi-analytical solution for pressure transient analysis of a hydraulically fractured vertical well in a bounded dual-porosity reservoir. *J. Hydrol.* **2018**, *565*, 289–301. [CrossRef]
9. Jiang, L.; Liu, J.; Liu, T.; Yang, D. A semianalytical model for transient pressure analysis of a horizontal well with non-uniform fracture geometry and shape-dependent conductivity in tight formations. *J. Pet. Sci. Eng.* **2020**, *195*, 107860. [CrossRef]
10. Fan, Z.; Parashar, R. Transient flow to a finite-radius well with wellbore storage and skin effect in a poroelastic confined aquifer. *Adv. Water Resour.* **2020**, *142*, 103604. [CrossRef]
11. Jahabani, A.; Aguilera, R. Well testing of tight gas reservoirs. *J. Can. Pet. Technol.* **2009**, *48*, 64–70. [CrossRef]
12. Zeng, B.; Cheng, L.; Li, C. Low velocity non-linear flow in ultra-low permeability reservoir. *J. Pet. Sci. Eng.* **2011**, *80*, 1–6. [CrossRef]
13. Zafar, A.; Su, Y.; Li, L.; Fu, J.; Mehmood, A.; Ouyang, W.; Zhang, M. Tight gas production model considering TPG as a function of pore pressure, permeability and water saturation. *Pet. Sci.* **2020**, *17*, 1356–1369. [CrossRef]
14. Wang, S.; Huang, Y.; Civan, F. Experimental and theoretical investigation of the Zaoyuan field heavy oil flow through porous media. *J. Pet. Sci. Eng.* **2006**, *50*, 83–101. [CrossRef]
15. Wei, X.; Qun, L.; Shusheng, G.; Zhiming, H.; Hui, X. Pseudo threshold pressure gradient to flow for low permeability reservoirs. *Pet. Explor. Dev.* **2009**, *36*, 232–236. [CrossRef]
16. Zhu, W.; Zou, G.; Liu, Y.; Liu, W.; Pan, B. The influence of movable water on the gas-phase threshold pressure gradient in tight gas reservoirs. *Energies* **2022**, *15*, 5309. [CrossRef]
17. Zhao, M.; Cao, M.; He, H.; Dai, C. Study on variation laws of fluid threshold pressure gradient in low permeable reservoir. *Energies* **2020**, *13*, 3704. [CrossRef]
18. Dou, H.; Ma, S.; Zou, C.; Yao, S. Threshold pressure gradient of fluid flow through multi-porous media in low and extra-low permeability reservoirs. *Sci. China Earth Sci.* **2014**, *57*, 2808–2818. [CrossRef]
19. Zhu, W.; Liu, Y.; Shi, Y.; Zou, G.; Zhang, Q.; Kong, D. Effect of dynamic threshold pressure gradient on production performance in water-bearing tight gas reservoir. *Adv. Geo-Energy Res.* **2022**, *6*, 286–295. [CrossRef]

20. Li, D.; Zha, W.; Liu, S.; Wang, L.; Lu, D. Pressure transient analysis of low permeability reservoir with pseudo threshold pressure gradient. *J. Pet. Sci. Eng.* **2016**, *147*, 308–316. [CrossRef]
21. Nie, R.; Wang, Y.; Kang, Y.; Jia, Y. Modeling the characteristics of Bingham porous-flow mechanics for a horizontal well in a heavy oil reservoir. *J. Pet. Sci. Eng.* **2018**, *171*, 71–81. [CrossRef]
22. Zhuang, H.; Sun, H.; Liu, X. *Dynamic Well Testing in Petroleum Exploration and Development*, 1st ed.; Elsevier: Oxford, UK, 2020.
23. Garcia-Rivera, J.; Raghavan, R. Analysis of short-time pressure data dominated by wellbore storage and skin. *J. Pet. Technol.* **1979**, *31*, 31–623. [CrossRef]
24. Wiewiorowski, N.; Valdes-Peres, A.; Blasingame, T. Characterization of Early-Time (Clean-Up) Performance for a Well with a Vertical Fracture Producing at Constant Pressure. In Proceedings of the Unconventional Resources Technology Conference, Austin, TX, USA, 24–26 July 2017. [CrossRef]
25. Wei, C.; Cheng, S.; Wang, X.; Li, W.; Li, Z.; Di, S.; Wen, C. A quick analysis approach for early-time well test data of vertically fractured wells: Applications in Changqing oilfield, China. *J. Pet. Sci. Eng.* **2021**, *201*, 108517. [CrossRef]
26. Stehfest, H. Numerical inversion of Laplace transforms algorithm 368. *Commun. ACM* **1970**, *13*, 47–49. [CrossRef]
27. Gringarten, A.C.; Bourdet, D.P.; Landel, P.A.; Kniazeff, V.J. A Comparison Between Different Skin and Wellbore Storage Type-Curves for Early-Time Transient Analysis. In Proceedings of the SPE Annual Technical Conference and Exhibition, Las Vegas, NV, USA, 23 September 1979. [CrossRef]
28. Ramey, H.J. Practical Use of Modern Well Test Analysis. In Proceedings of the SPE California Regional Meeting, Long Beach, CA, USA, 7 April 1976. [CrossRef]
29. Bourdet, D.; Whittle, T.; Douglas, A.; Pirard, Y. A new set of type curves simplifies well test analysis. *World Oil* **1983**, *196*, 95–106.
30. Lu, D.; Guo, Y.; Kong, X. A type curve for analyzing the early well testing data under unsteady flow. *Well Test.* **1993**, *2*, 33–40.
31. Lin, J.; He, H.; Han, Z. Typical curves and their analysis method for well test data without radial flow response. *Acta Pet. Sin.* **2017**, *38*, 562. [CrossRef]

MDPI AG
Grosspeteranlage 5
4052 Basel
Switzerland
Tel.: +41 61 683 77 34

Energies Editorial Office
E-mail: energies@mdpi.com
www.mdpi.com/journal/energies

www.ingramcontent.com/pod-product-compliance
Lightning Source LLC
Chambersburg PA
CBHW041349210926
51850CB00013B/3029
* 9 7 8 3 7 2 5 8 4 1 4 7 9 *